Studies in Systems, Decision and Control

Volume 644

The series "Studies in Systems, Decision and Control" (SSDC) covers both new developments and advances, as well as the state of the art, in the various areas of broadly perceived systems, decision making and control–quickly, up to date and with a high quality. The intent is to cover the theory, applications, and perspectives on the state of the art and future developments relevant to systems, decision making, control, complex processes and related areas, as embedded in the fields of engineering, computer science, physics, economics, social and life sciences, as well as the paradigms and methodologies behind them. The series contains monographs, textbooks, lecture notes and edited volumes in systems, decision making and control spanning the areas of Cyber-Physical Systems, Autonomous Systems, Sensor Networks, Control Systems, Energy Systems, Automotive Systems, Biological Systems, Vehicular Networking and Connected Vehicles, Aerospace Systems, Automation, Manufacturing, Smart Grids, Nonlinear Systems, Power Systems, Robotics, Social Systems, Economic Systems and other. Of particular value to both the contributors and the readership are the short publication timeframe and the worldwide distribution and exposure which enable both a wide and rapid dissemination of research output.

Indexed by SCOPUS, DBLP, WTI Frankfurt eG, zbMATH, SCImago.

All books published in the series are submitted for consideration in Web of Science.

Muammer Catak · Tofigh Allahviranloo

Discrete Random Processes

The Mathematics of Randomness

Springer

Muammer Catak
Electrical Engineering College of Engine
American University of the Middle East
Kuwait, Kuwait

Tofigh Allahviranloo
Research Center of Performance
and Productivity Analysis
Istinye University
Istanbul, Türkiye

ISSN 2198-4182 ISSN 2198-4190 (electronic)
Studies in Systems, Decision and Control
ISBN 978-3-032-08629-7 ISBN 978-3-032-08630-3 (eBook)
https://doi.org/10.1007/978-3-032-08630-3

This Springer imprint is published by the registered company Springer Nature Switzerland AG
The registered company address is: Gewerbestrasse 11, 6330 Cham, Switzerland

"Below is death
Above is hunger
Death is probable down below
Hunger is certain at above"

For all miners.

Preface

Overview

This textbook, entitled *Discrete Random Processes*, delivers a comprehensive and meticulously structured introduction to the essential concepts and techniques of signal processing and random processes. It is designed to bridge the gap between theoretical foundations and practical applications, with a particular emphasis on relevance to undergraduate and graduate studies in engineering and related disciplines.

The primary aim of this book is to equip students with a robust understanding of the fundamental principles underpinning digital signal processing and the analysis of random phenomena. It fosters the development of strong problem-solving skills and empowers readers to effectively apply theoretical knowledge to a wide range of engineering challenges. To achieve this, each chapter incorporates a variety of solved examples, ranging from introductory to advanced complexity, alongside a rich selection of problems that highlight the practical significance of the covered topics.

Chapter Descriptions

Chapter 1 provides a foundational review of Signals and Systems. It begins by defining the basic concepts of signals and systems and then delves into a detailed exploration of system properties, including linearity, time invariance, stability, causality, and memorylessness. The chapter concludes with methods for finding impulse responses of causal linear time-invariant (LTI) systems, setting the stage for subsequent analysis of system behavior.

Chapter 2 introduces the Z-transform, a powerful tool for analyzing discrete-time signals and systems. The chapter covers the crucial concept of the Region of Convergence (RoC), explores the properties of the Z-transform, such as linearity, time reversal, and the convolution sum, and examines the correlation of two real sequences. Finally, it discusses the system function of an LTI system, including poles and zeros, and the analysis of LTI systems using the system function.

Chapter 3 focuses on the Discrete-Time Fourier Transform (DTFT). It systematically explains the properties of the DTFT, including periodicity, linearity, time shift, frequency shift, differentiation in the frequency domain, the convolution sum in the time domain, multiplication in the time domain or convolution in the frequency domain, and modulation. The chapter also presents commonly used DTFT pairs, autocorrelation, and LTI system analysis using the DTFT.

Chapter 4 extends the discussion to the Discrete Fourier Transform(DFT) and its efficient implementation, the Fast Fourier Transform (FFT). It covers the implementation of the DFT, LTI systems analysis using the DFT, the FFT algorithm, sampling, 2D DFT algorithms, and applications of the FFT, including sinusoidal sequence analysis.

Chapters 5–7 transition to the probabilistic treatment of signals. Chapter 5 provides a thorough introduction to Random Variables, covering probability distribution functions (PDFs), probability density functions (pdfs), expected value and variance, common discrete probability distribution functions (Uniform, Binomial, Pascal, and Poisson), and the characteristic function. Chapter 6 introduces Random Processes, including averages, correlation functions, covariance functions, the Fourier transform in discrete-time signal processing, and the classification of random signals and processes. Chapter 7 delves into the important class of Discrete-Time Gaussian Random Processes, discussing their mathematical properties, including stationarity, correlation and covariance matrices, cross-correlation and cross-covariance, and the Fourier transform and spectral analysis (Power Spectral Density and Cross Power Spectral Density). It also explores engineering applications, such as Gaussian noise in digital communication, Kalman filtering, and financial modeling.

Chapters 8–10 cover important classes of random processes and their applications. Chapter 8 focuses on Markov Processes, covering fundamentals, building up a Markov process, classification of states in Markov chains, the random walk example, long-term behavior and limiting distribution, and applications in statistical signal processing, including queuing theory and image denoising. Chapter 9 is dedicated to Poisson Processes, covering basic definitions and related equations, properties, and applications in modeling website traffic, modeling particle counts, and modeling phone calls in a call center. Chapter 10 concludes the book with an introduction to the Ornstein-Uhlenbeck Process, its general definitions and related mathematical equations, and its application in financial mathematics, specifically in modeling short-term interest rates using the Vasicek model.

This book provides a rigorous yet accessible treatment of digital random processes, making it a valuable resource for students and practicing engineers seeking to master this essential field.

Kuwait, Kuwait
Istanbul, Türkiye
May 2025

Muammer Catak
Tofigh Allahviranloo

Contents

Chapter 1
Introduction to Signals and Systems

This chapter discusses the signals and systems in detail. Signal can be described as a mathematical function containing information about the system under investigation. Mathematical modelling has an significant role in almost all engineering applications. In order to analyse, to understand, and to control a physically realized system or a process, we usually notate some important variables as mathematical function, hence the process can be modelled by means of such functions. In Fig. 1.1, a simple block diagram representation of a PID control system with input and output signals is depicted.

Discrete-time signals can be represented by a mathematical functions which represent well-defined sequences, such as;

$$\text{The signal } \equiv x[n] \text{ where } n \text{ is an integer, and } -\infty < n < \infty \tag{1.1}$$

A representation of a discrete-time signal is illustrated in Fig. 1.2.

In order to improve our understanding of discrete-time signals, the unit impulse function, and the unit step function should be introduced.

The Unit Impulse Function

The unit impulse function, $\delta[n]$, is defined as;

$$\delta[n-k] = \begin{cases} 1 & \text{for } n = k \\ 0 & \text{elsewhere} \end{cases} \tag{1.2}$$

M. Catak and T. Allahviranloo, *Discrete Random Processes*, Studies in Systems, Decision and Control 644, https://doi.org/10.1007/978-3-032-08630-3_1

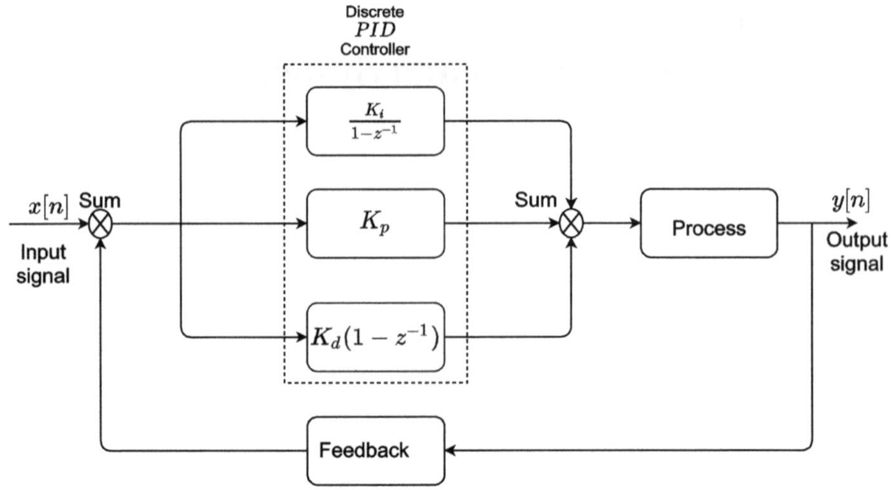

Fig. 1.1 A simple block diagram representation of a *PID* control system

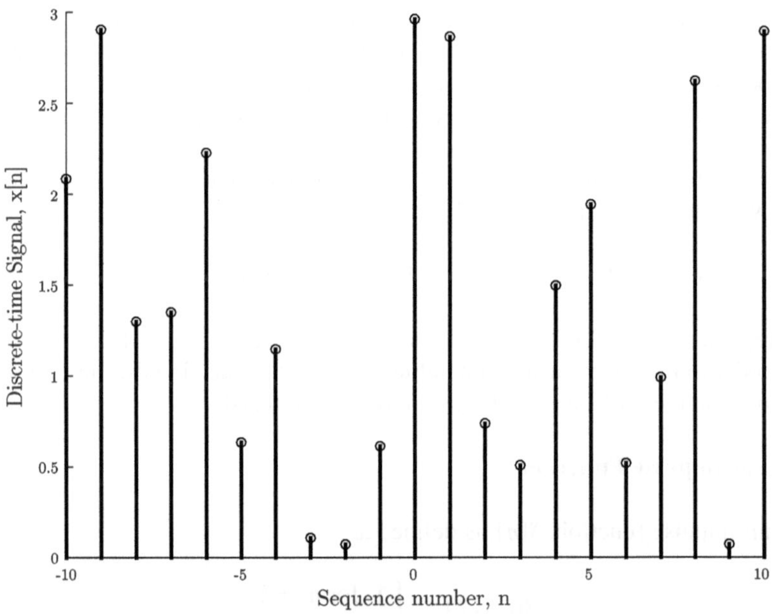

Fig. 1.2 A representation of a discrete-time signal

and

$$\sum_{-\infty}^{\infty} \delta[n-k] = 1 \quad \text{for a specific value of } k$$

In general, any discrete-time signal can be expressed as;

$$x[n] = \sum_{k=-\infty}^{\infty} x[k]\delta[n-k] \tag{1.3}$$

The Unit Step Function

The unit step function is defined as;

$$u[n-k] = \begin{cases} 1 & \text{for } n \geq k \\ 0 & \text{for } n < k \end{cases} \tag{1.4}$$

The relationship between the unit impulse function, and the unit step function is;

$$u[n] = \sum_{k=-\infty}^{n} \delta[k] \tag{1.5}$$

or, more conveniently,

$$u[n] = \sum_{k=0}^{\infty} \delta[n-k] \tag{1.6}$$

Illustrations of the unit impulse function and the unit step function is depicted in Fig. 1.3.

1.1 Signals and Systems

In signals and system theory, any system consists of three parts, such that the input denoted as $x[n]$ the impulse response $h[n]$ which characterizes the process, and the output $y[n]$ as shown in Fig. 1.4.

Let T be an operator ruling the input $x[n]$ to the output $y[n]$

$$y[n] = T\{x[n]\}$$

If the input is a unit impulse function, i.e. $\delta[n]$, then the corresponding output is called the impulse response; by definition

$$h[n] = T\{\delta[n]\}$$

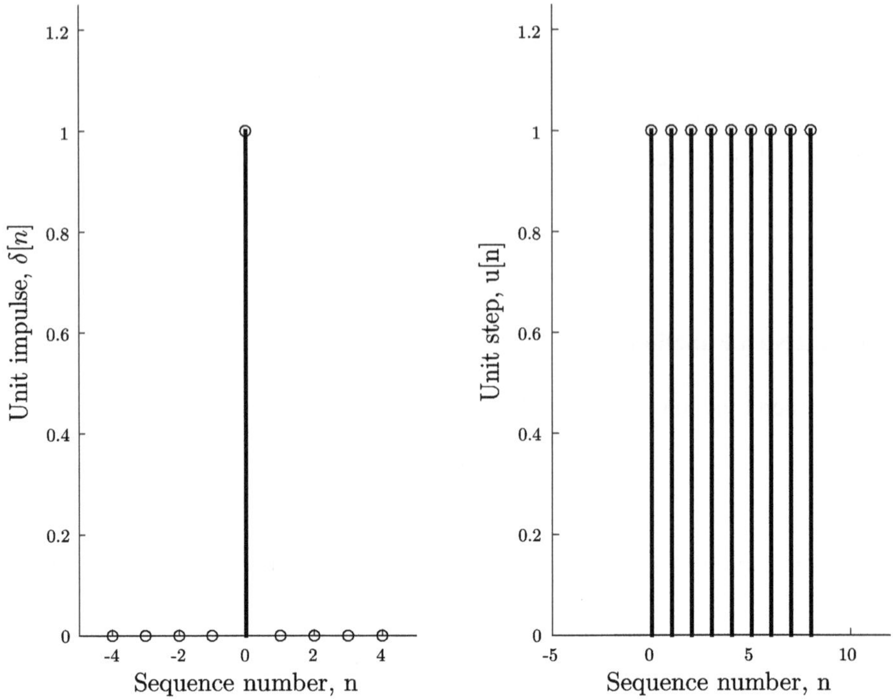

Fig. 1.3 Unit impulse function, and unit step function

Fig. 1.4 A basic system representation

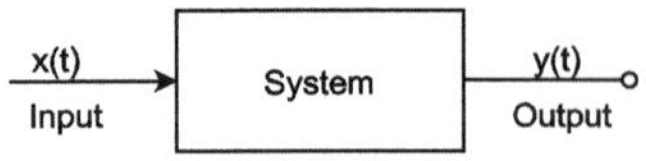

1.2 System Properties

1.2.1 Linear Systems

A system is called linear if and only if the ruling operator $T\{.\}$ has the following properties:

(i) $T\{\alpha x[n]\} = \alpha y[n]$, where α is a nonzero constant
(ii) $T\{x_1[n] + x_2[n]\} = T\{x_1[n]\} + T\{x_2[n]\}$

These are called *homogeneity* and *additivity* properties, respectively. Consequently, if these two properties generalized and combined into one definition;

$$y[n] = T\left\{\sum_{i=1}^{\infty} \alpha_i x_i[n]\right\} = \sum_{i=1}^{\infty} \alpha_i T\left\{x_i[n]\right\} = \sum_{i=1}^{\infty} \alpha_i y_i[n] \tag{1.7}$$

Example 1.1 Consider a system with the following input-output relation;

$$y[n] = 4x[n] + 3$$

Determine whether the system is linear or not.

Solution 1.1

Let assign two arbitrary input functions as $x_1[n]$ and $x_2[n]$

$$y_1[n] = 4x_1[n] + 3$$

$$y_2[n] = 4x_2[n] + 3$$

$$y[n] = 4(x_1[n] + x_2[n]) + 1$$

$$y_1[n] + y_2[n] = 4x_1[n] + 4x_2[n] + 6$$

$y[n] \neq y_1[n] + y_2[n]$ therefore, the system is nonlinear.

1.2.2 Time-Invariant Systems

A system is time-invariant if any time shift in the input signal results in the same time shift in the output signal, such that;

$$\begin{aligned} y[n] &= T\{x[n])\} \\ y\left[n - n_0\right] &= T\left\{x\left[n - n_0\right]\right\} \end{aligned} \tag{1.8}$$

Otherwise, the system is called time-varying.

Example 1.2 Consider a system with the given input-output relationship;

$$y[n] = 3x[n] - 2$$

Determine whether the system is time-invariant or not.

Solution 1.2

$$y[n-n_0] = T\{x[n-n_0])\}$$
$$= 3x[n-n_0] - 2$$

which means that if the input is shifted in time with n_0, then the corresponding output will be shifted in time with n_0. Therefore, the system is time-invariant.

1.2.3 Stable Systems

In a general point of view, if a system returns its equilibrium points after any interruption by external source, such system is called a stable system.

If the input of a system is bounded, $|x[n]| < K$, where K is a finite number; the corresponding output $y[n]$ is also bounded if and only if the impulse response of the system $h[n]$ satisfies the following condition;

$$\sum_{-\infty}^{\infty} |h[n]| < \infty \tag{1.9}$$

Such system is called bounded input bounded output (BIBO) stable.

Example 1.3 Consider a system described as;

$$y[n] = 2x[n] - x[n-1]$$

Determine whether the system is stable or unstable.

Solution 1.3

If the input is assign as unit impulse function $\delta[n]$, then the corresponding output will be the impulse response, $h[n]$;

$$h[n] = 2\delta[n] - \delta[n-1]$$

it will be checked whether the following statement is hold or not;

$$\sum_{-\infty}^{\infty} |h[n]| < \infty$$

$$\sum_{-\infty}^{\infty} |2\delta[n] - \delta[n-1]| = 1 < \infty$$

therefore, the system is stable.

1.2.4 Causal Systems

A system is said to be causal if the output of the system $y[n]$ at an arbitrary time n_0 depends only on the input at and or before the time n_0. In other words, the present of output can depend on present and/or past inputs. This definition is hold only if the impulse response of the system $h[n]$ satisfies;

$$h[n] = 0 \quad \forall \quad n < 0 \tag{1.10}$$

In literature, the causal systems are also called as physically realizable systems.

Example 1.4 Consider a system described as;

$$y[n] = x[n] + 2x[n-1]$$

Determine whether the system is causal or noncausal.

Solution 1.4

Let select an arbitrary time as n_0;

$$y[n_0] = x[n_0] + 2x[n_0 - 1]$$

since the output is only depended on $n \leq n_0$, then the system is called causal.

1.2.5 Linear Time-Invariant Systems

In this section, the expression of the output of a LTI system in terms of the input $x[n]$ and the impulse response $h[n]$ will be discussed. Any discrete-time function can be represented by using unit impulse function;

$$x[n] = \sum_{k=-\infty}^{\infty} x[k]\delta[n-k] \tag{1.11}$$

Thereafter, the output of such a system can be defined as;

$$\begin{aligned} y[n] &= T\{x[n])\} \\ &= T\left\{\sum_{k=-\infty}^{\infty} x[k]\delta[n-k]\right\} \\ &= \sum_{k=-\infty}^{\infty} x[k]T\{\delta[n-k]\} \end{aligned} \tag{1.12}$$

If the system is linear (no information about the time-invariance of the system), then the impulse response can be expressed as;

$$T\{\delta[n-k]\} = h[n,k]$$

hence, the output $y[n]$ can be derived as;

$$y[n] = \sum_{k=-\infty}^{\infty} x[k]h[n,k] \tag{1.13}$$

which is called superposition theorem in signal processing.

If the system is linear and time-invariant (LTI), then

$$T\{\delta[n-k]\} = h[n-k]$$

thereafter, the output signal is represented as;

$$y[n] = \sum_{k=-\infty}^{\infty} x[k]h[n-k] \tag{1.14}$$

which is called the convolution sum, and it will be denoted as follows in this book

$$y[n] = x[n] * h[n] \tag{1.15}$$

Example 1.5 Consider a system described as;

$$y[n] = x[n] + \frac{1}{2}x[n-1] + \frac{1}{4}x[n-2]$$

Determine whether the system is LTI or not.

Solution 1.5

Let $x_1[n]$ and $x_2[n]$ be two arbitrary inputs

$$y_1[n] = x_1[n] + \frac{1}{2}x_1[n-1] + \frac{1}{4}x_1[n-2]$$

$$y_2[n] = x_2[n] + \frac{1}{2}x_2[n-1] + \frac{1}{4}x_2[n-2]$$

If the input $x[n] = x_1[n] + x_2[n]$, then

$$y[n] = (x_1[n] + x_2[n]) + \frac{1}{2}(x_1[n-1]x_2[n-1]) + \frac{1}{4}(x_1[n-2] + x_2[n-2])$$

$$y[n] = y_1[n] + y_2[n]$$

then, the system is linear.
If an arbitrary time n_0 is assigned;

$$y[n_0] = x[n - n_0] + \frac{1}{2}x[(n - n_0) - 1] + \frac{1}{4}x[(n - n_0) - 2]$$

hence, the system is time-invariant. Therefore, the system is an LTI system.

1.2.6 Memoryless Systems

A system is called memoryless if the output $y[n$ is determined by the input $x[n]$ at the same value of any arbitrary n. Otherwise, the systems is labelled as having memory. This definition covers not only time-delay case but also time-advance cases as well.

Example 1.6 Determine the following systems are memoryless or having memory;

(a) $y[n] = 2x[n] - 3$
(b) $y[n] = x[n-3]$
(c) $y[n] = 2x[n] - x[n+1]$

Solution 1.6

(a) Since $y[n]$ depends merely on $x[n]$ at any arbitrary value of n, therefore this system is memoryless.
(b) Since $y[n_0]$ depends on $x[n_0 - 3]$ (time-delay) at any arbitrary value of n_0, therefore this system has memory.
(c) $y[n_0]$ depends on $x[n_0 + 1]$ (time-advance) at any arbitrary value of n_0, therefore this system has memory.

1.2.7 Finding Impulse Responses of Causal LTI Systems

A common type of LTI systems are represented by Nth order constant coefficient difference equations. In general, the relation between the input and the output is given as;

$$\sum_{k=0}^{N} \alpha_k y[n-k] = \sum_{m=0}^{M} \beta_k n[n-m] =$$

For such equations, the impulse response can be derived as invoking $\delta[n]$ instead of $x[n]$, then the corresponding output will be the impulse response by definition.

Example 1.7 Derive the impulse response, $h[n]$, of the following causal LTI system given as;

$$y[n] - \frac{1}{2}y[n-1] = x[n] - x[n-1]$$

Solution 1.7

Since the given system is an LTI system, if the input is $\delta[n]$, then the corresponding output will be $h[n]$

$$h[n] - \frac{1}{2}h[n-1] = \delta[n] - \delta[n-1]$$

Considering a causal system, the impulse response must satisfy

$$h[n] = 0 \quad \forall \quad n < 0$$

then,

$$h[n] = \frac{1}{2}h[n-1] + \delta[n] - \delta[n-1]$$

$$\begin{aligned} h[0] &= \frac{1}{2}h[-1] + \delta[0] - \delta[-1] \\ &= 0 + 1 - 0 \\ &= 1 \end{aligned}$$

$$\begin{aligned} h[1] &= \frac{1}{2}h[0] + \delta[1] - \delta[0] \\ &= \frac{1}{2} + 0 - 1 \\ &= -\frac{1}{2} \end{aligned}$$

$$
\begin{aligned}
h[2] &= \frac{1}{2}h[1] + \delta[2] - \delta[1] \\
&= -\frac{1}{4} + 0 - 0 \\
&= -\frac{1}{4} \\
h[3] &= \frac{1}{2}h[2] + \delta[3] - \delta[2] \\
&= -\frac{1}{8} + 0 - 0 \\
&= -\frac{1}{8}
\end{aligned}
$$

In general, the impulse response can be written as;

$$h[n] = -\left(\frac{1}{2}\right)^n u[n-1] + \delta[n]$$

Example 1.8 Consider a causal LTI system given as;

$$y[n] - y[n-1] = x[n]$$

(a) Derive the impulse response, $h[n]$
(b) Find the corresponding output if $x[n] = \left(\frac{1}{2}\right)^n u[n]$

Solution 1.8

(a) Assuming with the input is $\delta[n]$, then the output will be $h[n]$

$$h[n] = h[n-1] + \delta[n]$$

$$
\begin{aligned}
h[0] &= h[-1] + \delta[0] \\
&= 0 + 1 \\
&= 1
\end{aligned}
$$

$$
\begin{aligned}
h[1] &= h[0] + \delta[1] \\
&= 1 + 0 \\
&= 1
\end{aligned}
$$

$$h[2] = h[1] + \delta[2]$$
$$= 1 + 0$$
$$= 1$$
$$\dots$$

Therefore, the impulse response can be written as;

$$h[n] = u[n]$$

(b) Since $y[n] = x[n] * h[n]$

$$y[n] = \sum_{k=-\infty}^{\infty} x[k]h[n-k]$$
$$= \sum_{k=-\infty}^{\infty} \left(\frac{1}{2}\right)^k u[k]u[n-k]$$

if $k < 0$, then $u[k] = 0$

$$= \sum_{k=0}^{\infty} \left(\frac{1}{2}\right)^k u[k]u[n-k]$$

if $k > n$, then $u[n-k] = 0$

$$= \sum_{k=0}^{n} \left(\frac{1}{2}\right)^k u[k]u[n-k]$$
$$= \sum_{k=0}^{n} \left(\frac{1}{2}\right)^k$$

using sum of geometric series property

$$y[n] = \frac{1 - \left(\frac{1}{2}\right)^{n+1}}{1 - \frac{1}{2}}, \quad n \geq 0$$

or

$$y[n] = \frac{1 - \left(\frac{1}{2}\right)^{n+1}}{1 - \frac{1}{2}} u[n]$$

The input, impulse response, and the corresponding output are shown in Fig. 1.5.

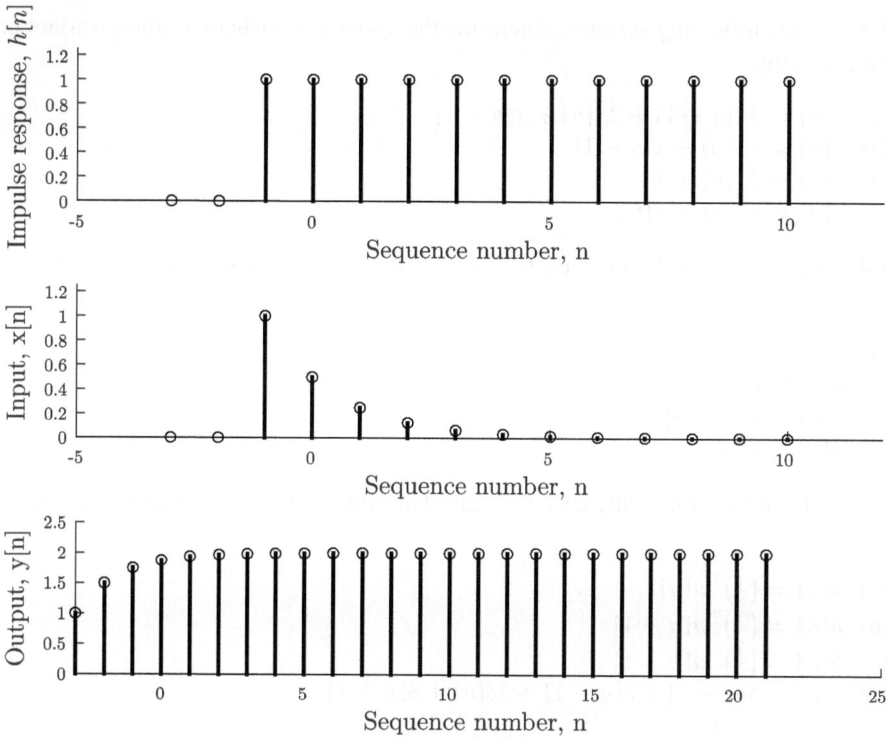

Fig. 1.5 The input, impulse response, and the corresponding output of Example 1.8

Problems

1.1 Determine the following systems are whether linear or not.

(a) $y[n] = 2x[n] - 3$
(b) $y[n] = n^2 x[n]$
(c) $y[n] = 2x[n] - x[n-1]$
(d) $y[n] = 2e^{x[n]}$

1.2 For the following systems, determine the systems are whether causal or not.

(a) $y[n] = x[n] - \frac{1}{2}x[n-1] - \frac{1}{4}x[n-2]$
(b) $y[n] = \frac{1}{3}(x[n+1] + x[n] + x[n-1])$
(c) $y[n] = 2e^{x[n]} - x[n]$
(d) $y[n] = (n-2)x[n]$

1.3 For the following systems, determine the systems are whether time-invariant or time-varying.

(a) $y[n] = 3x[n-1] + 2x[n] + x[n-1]$
(b) $y[n] = nx[n] - x[n-1]$
(c) $y[n] = 2x[n] + 1$
(d) $y[n] = x[2n] - x[n]$

1.4 Consider the following each systems, determine the systems are stable or unstable.

(a) $h[n] = \left(\frac{1}{2}\right)^n u[n]$
(b) $h[n] = u[n]$
(c) $h[n] = \left(-\frac{1}{2}\right)^n u[n]$
(d) $h[n] = 2^n u[n]$

1.5 Consider the following each systems, determine the systems whether are causal or not.

(a) $h[n] = \left(\frac{1}{2}\right)^n u[n]$
(b) $h[n] = \left(\frac{1}{2}\right)^n u[n-1]$
(c) $h[n] = \left(\frac{1}{2}\right)^n u[n+1]$
(d) $h[n] = \delta[n-2] + \delta[n-1] + 3\delta[n] + \delta[n+1]$

1.6 Consider the following causal LTI system described as

$$y[n] + 2y[n-1] = x[n] + x[n-1]$$

(a) Derive the impulse response, $h[n]$
(b) Calculate the output if the input $x[n] = u[n-1] - u[n-5]$

1.7 Consider the following causal LTI system described as

$$y[n] + 2y[n-1] - y[n-2] = x[n] + x[n-1]$$

(a) Derive the impulse response, $h[n]$
(b) Calculate the output if the input $x[n] = 2u[n] + \delta[n-1]$

1.8 Calculate the output of an LTI system if the impulse response, $h[n]$, and the input, $x[n]$ are

$$h[n] = \frac{1}{4}u[n]$$

$$x[n] = u[n-1] - u[n+5]$$

1.9 Determine the following systems are whether linear or not.

(a) $y[n] = x^2[n]$
(b) $y[n] = \sin(x[n])$
(c) $y[n] = x[n]u[n]$ (where $u[n]$ is the unit step function)
(d) $y[n] = \sum_{k=-\infty}^{n} x[k]$

1.10 For the following systems, determine the systems are whether causal or not.

(a) $y[n] = x[n] + x[n-3] + x[n-5]$
(b) $y[n] = x[n-1] + x[n+2]$
(c) $y[n] = \cos(x[n])$
(d) $y[n] = \sum_{k=-\infty}^{n+1} x[k]$

1.11 For the following systems, determine the systems are whether time-invariant or time-varying.

(a) $y[n] = x[n] + x[-n]$
(b) $y[n] = x[n]\cos(\omega n)$ (where ω is a constant)
(c) $y[n] = e^{x[n]}$
(d) $y[n] = \sum_{k=n-2}^{n+2} x[k]$

1.12 Consider the following each systems, determine the systems are stable or unstable.

(a) $h[n] = \cos(n)u[n]$
(b) $h[n] = e^{-2n}u[n]$
(c) $h[n] = \delta[n] + 2\delta[n-1]$
(d) $h[n] = nu[n]$

1.13 Consider the following each systems, determine the systems whether are causal or not.

(a) $h[n] = u[n] - u[n-5]$
(b) $h[n] = \delta[n+5]$
(c) $h[n] = n(u[n] - u[n-4])$
(d) $h[n] = \sum_{k=0}^{\infty} \left(\frac{1}{3}\right)^k \delta[n-k]$

1.14 Consider the following causal LTI system described as

$$y[n] - \frac{1}{2}y[n-1] = x[n]$$

(a) Derive the impulse response, $h[n]$
(b) Calculate the output if the input $x[n] = \delta[n] - \delta[n-1]$

1.15 Consider the following causal LTI system described as

$$y[n] - \frac{1}{3}y[n-1] - \frac{1}{4}y[n-2] = x[n]$$

(a) Derive the impulse response, $h[n]$ (Assume distinct roots in the characteristic equation)
(b) Calculate the output if the input $x[n] = u[n]$

1.16 Calculate the output of an LTI system if the impulse response, $h[n]$, and the input, $x[n]$ are

$$h[n] = \delta[n] - \delta[n-1]$$
$$x[n] = 2^n u[-n]$$

1.17 A discrete-time LTI system has the impulse response

$$h[n] = \begin{cases} 1, & 0 \le n \le 2 \\ 0, & \text{otherwise} \end{cases}$$

Determine the output $y[n]$ when the input is

$$x[n] = \begin{cases} 1, & 0 \le n \le 2 \\ 0, & \text{otherwise} \end{cases}$$

1.18 Determine the output $y[n]$ of a discrete-time LTI system with impulse response

$$h[n] = \left(\frac{1}{2}\right)^n u[n]$$

when the input is

$$x[n] = u[n] - u[n-4]$$

1.19 An LTI system is described by the difference equation

$$y[n] - \frac{1}{4}y[n-1] = x[n] + \frac{1}{8}x[n-1]$$

(a) Find the impulse response of the system.
(b) Determine if the system is stable.

1.20 Two LTI systems with impulse responses $h_1[n]$ and $h_2[n]$ are connected in cascade.

$$h_1[n] = u[n] - u[n-2]$$

$$h_2[n] = \delta[n] + \delta[n-1]$$

(a) Find the overall impulse response of the cascaded system.
(b) If the input to the cascaded system is $x[n] = \delta[n] - \delta[n-1]$, find the output.

Chapter 2
Linear Transformations: Z-Transform

Z-transform is one of the powerful tools in discrete-time signal processing to analyse those systems, and to investigate the input-output relationships. This chapter is dedicated for Z-transform and its applications in signal processing area.

The Z-transform is a linear operator, which is notated as $\mathcal{Z}\{x[n]\}$. It transforms a discrete-time signal, $x[n]$, into a continuous complex domain, which is called z-domain. The transformation operates as;

$$X(z) = \sum_{n=-\infty}^{\infty} x[n]z^{-n} \tag{2.1}$$

where $x[n]$ and $X(z)$ are known as Z-transform pairs.

2.1 Region of Convergence (RoC)

Z-transform of a discrete-time signal may not be exist. The condition to have Z-transform of a discrete-time signal, $x[n]$, i.e. to be able to derive $X(z)$, is given as;

$$\begin{aligned} &\sum_{n=-\infty}^{\infty} |x[n]z^{-n}| < \infty \\ &= \sum_{n=-\infty}^{\infty} |x[n]||z^{-n}| < \infty \end{aligned} \tag{2.2}$$

where $|z|$ represents the magnitude of $z = re^{jw}$.

M. Catak and T. Allahviranloo, *Discrete Random Processes*, Studies in Systems, Decision and Control 644, https://doi.org/10.1007/978-3-032-08630-3_2

The values of z, where the Z-transform converges, and then $X(z)$ can be obtained, is called region of convergence (RoC).

Z-transform of $x[n] = \alpha^n u[n]$

$$X(z) = \sum_{n=-\infty}^{\infty} (\alpha)^n u[n] z^{-n}$$
$$= \sum_{n=0}^{\infty} \left(\alpha z^{-1}\right)^n$$

Using the geometric summation formula

$$\sum_{n=0}^{\infty} r^n = \frac{1}{1-r}, \quad |r| < 1$$

thereafter

$$X(z) = \frac{1}{1-\alpha z^{-1}}, \quad RoC := |z| > |\alpha|$$

Z-transform of $x[n] = -(\alpha)^n u[-n-1]$

$$X(z) = -\sum_{n=-\infty}^{+\infty} \alpha^n u[-n-1] z^{-n} = -\sum_{n=-\infty}^{-1} \alpha^n z^{-n}$$
$$= -\sum_{n=1}^{\infty} \alpha^{-n} z^n = 1 - \sum_{n=0}^{\infty} \left(\alpha^{-1} z\right)^n$$
$$X(z) = 1 - \frac{1}{1-a^{-1}z}$$
$$= \frac{1}{1-\alpha z^{-1}}, \quad RoC := |z| < |\alpha|$$

Z-transform of a Finite Sequence

Let a finite length of discrete time signal be defined as;

$$x(n) = \begin{cases} \alpha^n & 0 \le n \le N-1 \\ 0 & \text{otherwise} \end{cases}$$

It can be rewritten including the unit step function, $u[n]$;

$$x(n) = \alpha^n (u(n) - u(n-N))$$

The corresponding Z-transform can be calculated as;

$$\begin{aligned} X(z) &= \sum_{n=0}^{N-1} \alpha^n z^{-n} \\ &= \sum_{n=0}^{N-1} \left(\alpha z^{-1}\right)^n \\ &= \frac{1-\left(\alpha z^{-1}\right)^N}{1-\alpha z^{-1}} \\ &= \frac{1}{z^{N-1}} \frac{z^N - \alpha^N}{z-\alpha} \end{aligned}$$

While αz^{-1} is finite, the sum definitely converges due to the limited amount of nonzero elements. The RoC is the whole z-plane aside from the origin $z = 0$, in which there are no constraints on α as long as $|\alpha| < \infty$, i.e. $RoC = Z - \{0\}$.

Example 2.1 Derive the Z-transform of the following discrete-time signal, check the RoC.

$$x[n] = u[n]$$

Solution 2.1

$$\begin{aligned} X(z) &= \sum_{n=-\infty}^{\infty} x[n] z^{-n} \\ &= \sum_{n=-\infty}^{\infty} u[n] z^{-n} \\ &= \sum_{n=0}^{\infty} z^{-n} \\ &= \sum_{n=0}^{\infty} \left(\frac{1}{z}\right)^n \end{aligned}$$

considering the convergence of the power series, in order to have $X(z)$, $|\frac{1}{z}| < 1$ or $|z| > 1$, then

$$X(z) = \frac{1}{1-z^{-1}}, \quad \text{RoC:=}|z| > 1$$

The corresponding RoC is shown in Fig. 2.1.

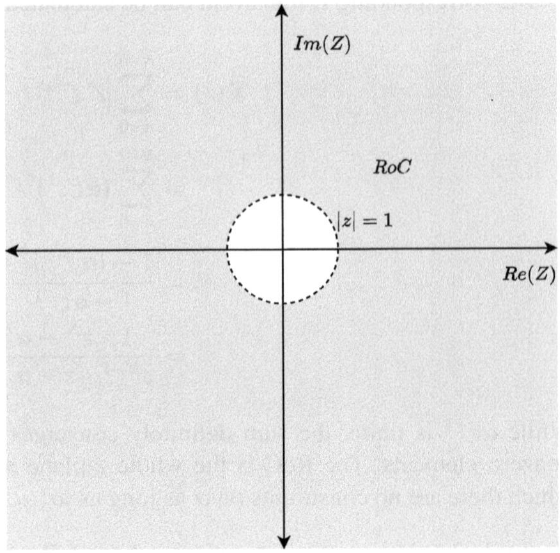

Fig. 2.1 Region of convergence for Example 2.1

Example 2.2 Calculate the Z-transform of the following discrete-time signal, discuss about the RoC.

$$x[n] = \left(\frac{1}{2}\right)^n u[n] - \left(\frac{1}{4}\right)^n u[n]$$

Solution 2.2

$$X(z) = \sum_{n=-\infty}^{\infty} x[n]z^{-n}$$

$$= \sum_{n=-\infty}^{\infty} \left(\left(\frac{1}{2}\right)^n u[n] - \left(\frac{1}{4}\right)^n u[n]\right) z^{-n}$$

$$= \sum_{n=0}^{\infty} \left(\left(\frac{1}{2}\right)^n - \left(\frac{1}{4}\right)^n\right) z^{-n}$$

$$= \sum_{n=0}^{\infty} \left(\frac{1}{2}\right)^n z^{-n} - \left(\frac{1}{4}\right)^n z^{-n}$$

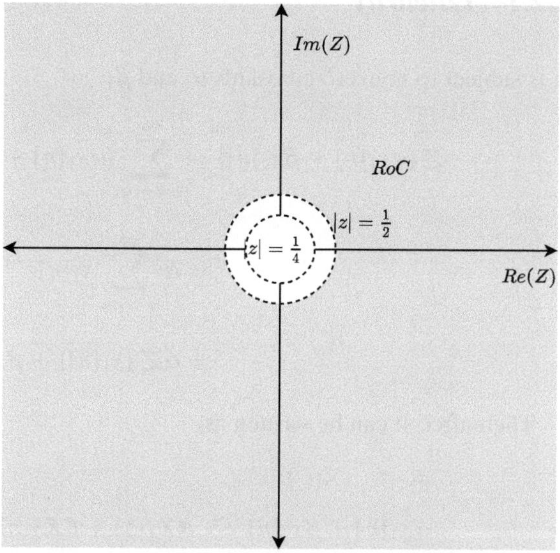

Fig. 2.2 Region of convergence for Example 2.2

considering the convergence of the power series, in order to have $X(z)$, $|z| > \frac{1}{2}$, and $|z| > \frac{1}{4}$. The intersection of these two conditions defines the RoC of the transformation. Therefore,

$$X(z) = \frac{1}{1 - \frac{1}{2}z^{-1}} - \frac{1}{1 - \frac{1}{4}z^{-1}}, \quad \text{RoC:=}|z| > \frac{1}{2}$$

The corresponding RoC is shown in Fig. 2.2.

2.2 Properties of Z-Transform

In this section, properties of Z-transform will be discussed under the following assumptions;

$$\begin{aligned} \mathcal{Z}\{x[n]\} &= X(z), \quad \text{RoC:=}R \\ \mathcal{Z}\{x_1[n]\} &= X_1(z), \quad \text{RoC:=}R_1 \\ \mathcal{Z}\{x_2[n]\} &= X_2(z), \quad \text{RoC:=}R_2 \end{aligned} \tag{2.3}$$

2.2.1 Linearity

It is subject to nonzero constants α, and β;

$$
\begin{aligned}
\mathcal{Z}\{\alpha x_1[n] + \beta x_2[n]\} &= \sum_{n=-\infty}^{\infty} (\alpha x_1[n] + \beta x_2[n])\, z^{-n} \\
&= \alpha \sum_{n=-\infty}^{\infty} x_1[n] z^{-n} + \beta \sum_{n=-\infty}^{\infty} x_2[n] z^{-n} \\
&= \alpha \mathcal{Z}\{x_1[n]\} + \beta \mathcal{Z}\{x_2[n]\}
\end{aligned}
\tag{2.4}
$$

Thereafter, it can be written as;

$$\alpha x_1[n] + \beta x_2[n] \overset{\mathcal{Z}}{\longleftrightarrow} \alpha X_1(z) + \beta X_2(z), \quad \text{RoC:=} R_1 \cap R_2 \tag{2.5}$$

2.2.1.1 Time Shifting

Considering an arbitrary constant n_0;

$$\mathcal{Z}\{x[n-n_0]]\} = \sum_{n=-\infty}^{\infty} x[n-n_0] z^{-n}$$

with the substitution of $k = n - n_0$

$$
\begin{aligned}
&= \sum_{k=-\infty}^{\infty} x[k] z^{-(k+n_0)} \\
&= z^{-n_0} \sum_{k=-\infty}^{\infty} x[k] z^{-k} \\
&= z^{-n_0} \mathcal{Z}\{x[n]\}
\end{aligned}
\tag{2.6}
$$

Hence, it is given as;

$$x[n-n_0] \overset{\mathcal{Z}}{\longleftrightarrow} z^{-n_0}X(z), \quad \text{RoC:=}R \tag{2.7}$$

2.2.2 Multiplication with an Exponential Sequence

Considering an arbitrary z_0;

$$\begin{aligned}
\mathcal{Z}\left\{z_0^n x[n]\right\} &= \sum_{n=-\infty}^{\infty} z_0^n x[n] z^{-n} \\
&= \sum_{n=-\infty}^{\infty} x[n] \left(\frac{z}{z_0}\right)^{-n} \\
&\text{with the substitution of } k = \frac{z}{z_0} \\
&= \sum_{n=-\infty}^{\infty} x[n] k^{-n} \\
&= X(k) \\
&= X\left(\frac{z}{z_0}\right)
\end{aligned} \tag{2.8}$$

Accordingly, it can be stated that;

$$z_0^n x[n] \overset{\mathcal{Z}}{\longleftrightarrow} X\left(\frac{z}{z_0}\right), \quad \text{RoC:=} |z_0| R \tag{2.9}$$

2.2.2.1 Multiplication by *n*, or Differentiation in *z*

The Z-transform of $nx[n]$ can be obtained as;

$$\begin{aligned}\mathcal{Z}\{nx[n]\} &= \sum_{n=-\infty}^{\infty} nx[n]z^{-n} \\ &= -z \sum_{n=-\infty}^{\infty} (-n)x[n]z^{-n-1} \\ &= -z\frac{\mathrm{d}X(z)}{\mathrm{d}z}\end{aligned} \tag{2.10}$$

thus, it can be written as;

$$nx[n] \overset{\mathcal{Z}}{\longleftrightarrow} -z\frac{\mathrm{d}X(z)}{\mathrm{d})z}, \quad \text{RoC:}= R \tag{2.11}$$

2.2.3 *Time Reversal*

The Z-transform of a time-reversal signal $x[-n]$ can be derived as;

$$\begin{aligned}\mathcal{Z}\{x[-n]\} &= \sum_{n=-\infty}^{\infty} x[-n]z^{-n} \\ &= \sum_{n=-\infty}^{\infty} x[n]z^{n} \\ &= \sum_{n=-\infty}^{\infty} x[n]\left(\frac{1}{z}\right)^{-n} \\ &= X\left(\frac{1}{z}\right)\end{aligned} \tag{2.12}$$

then, it can be given as;

$$x[-n] \overset{\mathcal{Z}}{\leftrightarrow} X\left(\frac{1}{z}\right), \quad \text{RoC:}= \frac{1}{R} \tag{2.13}$$

2.2.4 Convolution Sum

Let assume $x_1[n]$, and $x_2[n]$ are discrete signals and $y[n]$ is defined as convolution sum of these signals;

$$\begin{aligned} y[n] &= x_1[n] * x_2[n] \\ &= \sum_{k=-\infty}^{\infty} x_1[k]x_2[n-k] \end{aligned} \tag{2.14}$$

$$\begin{aligned} \mathcal{Z}\{y[n]\} &= \sum_{n=-\infty}^{\infty} y[n]z^{-n} \\ &= \sum_{n=-\infty}^{\infty} \left\{ \sum_{k=-\infty}^{\infty} x_1[k]x_2[n-k] \right\} z^{-n} \\ &= \underbrace{\sum_{k=-\infty}^{\infty} x_1[k]z^{-k}}_{X_1(z)} \underbrace{\sum_{m=-\infty}^{\infty} x_2[m]z^{-m}}_{X_2(z)} \end{aligned} \tag{2.15}$$

therefore, Z-transform of convolution sum of two signals is;

$$x_1[n] * x_2[n] \overset{\mathcal{Z}}{\leftrightarrow} X_1(z)X_2(z), \quad \text{RoC:}= R_1 \cap R_2 \tag{2.16}$$

2.2.5 Correlation of Two Real Sequences

Considering $x[n]$, and $y[n]$ are discrete real-valued sequences, the correlation between $x[n]$, and $x[n]$ can be calculated as;

$$r_{xy}[n] = \sum_{k=-\infty}^{\infty} x[k]y[n+k]$$

which equals

$$= x[n] * y[-n] \tag{2.17}$$

Applying Z-transform both sides of the equation, and using the time-reversal property of the Z-transform;

$$\mathcal{Z}\left\{r_{xy}[n]\right\} = \mathcal{Z}\left\{x[n] * y[-n]\right\}$$

$$R_{XY}(z) = X(z)Y\left(\frac{1}{z}\right) \tag{2.18}$$

Some commonly used Z-transform pairs are shown in Table 2.1.
Some important properties of Z-transform are depicted in Table 2.2.

Table 2.1 Some commonly used Z-transform pairs

Signal	Z-transform	*RoC*
$\delta[n]$	1	All z
$u[n]$	$\frac{1}{1-z^{-1}}$	$\|z\| > 1$
$nu[n]$	$\frac{z^{-1}}{(1-z^{-1})^2}$	$\|z\| > 1$
$(\alpha)^n u[n]$	$\frac{1}{1-\alpha z^{-1}}$	$\|z\| > \|\alpha\|$
$n(\alpha)^n u[n]$	$\frac{\alpha z^{-1}}{(1-\alpha z^{-1})^2}$	$\|z\| > \|\alpha\|$
$-(\alpha)^n u[-n-1]$	$\frac{1}{1-\alpha z^{-1}}$	$\|z\| < \|\alpha\|$
$-n(\alpha)^n u[-n-1]$	$\frac{\alpha z^{-1}}{(1-\alpha z^{-1})^2}$	$\|z\| < \|\alpha\|$
$(\cos \omega_0 n)\, u[n]$	$\frac{1-z^{-1}\cos\omega_o}{1-2z^{-1}\cos\omega_0+z^{-2}}$	$\|z\| > 1$
$(\sin \omega_0 n)\, u[n]$	$\frac{z^{-1}\sin\omega_o}{1-2z^{-1}\cos\omega_o+z^{-2}}$	$\|z\| > 1$

Table 2.2 Some commonly used properties of Z-transform

Property	Sequence	Transform	ROC
	$x[n]$	$X(z)$	R
	$x_1[n]$	$X_1(z)$	R_1
	$x_2[n]$	$X_2(z)$	R_2
Linearity	$\alpha x_1[n] + \beta x_2[n]$	$\alpha X_1(z) + \beta X_2(z)$	$R_1 \cap R_2$
Time-shift	$x[n - n_0]$	$z^{-n_0} X(z)$	R_x
Exponential sequence	$z_0^n x[n]$	$X\left(\frac{z}{z_0}\right)$	$\|z_0\| R$
Differentiation in z	$nx[n]$	$-z\frac{dX(z)}{dz}$	R
Time reversal	$x[-n]$	$X\left(\frac{1}{z}\right)$	$1/R$
Convolution Sum	$x_1[n] * x_2[n]$	$X_1(z)X_2(z)$	$R_1 \cap R_2$
Correlation	$r_{xy}(n) = x[n] * y[-n]$	$X(z)Y\left(\frac{1}{z}\right)$	$R_x \cap \frac{1}{R_y}$

2.3 The System Function of an LTI System, $H(z)$

Considering an LT systems, the output can be expressed by means of convolution sum of the input and the corresponding impulse response of the system;

$$y[n] = x[n] * h[n]$$

Therefore, Z-transform of the output can be written as using the convolution property given in Table 2.2;

$$Y(z) = X(z)H(z) \tag{2.19}$$

where $H(z)$ is the Z-transform of the impulse response, $h[n]$, and it is called *System Function* of the related LTI system.

Example 2.3 Derive the Z-transform of the following discrete-time signal, discuss about its RoC.

$$x[n] = \left(\frac{1}{2}\right)^n u[n] - \left(\frac{1}{8}\right)^n u[-n-1]$$

Solution 2.3 Referring to Table 2.1

$$x_1[n] = \left(\frac{1}{2}\right)^n u[n] \rightarrow RoC := R_1 = |z| > \frac{1}{2}$$

and

$$x_2[n] = -\left(\frac{1}{8}\right)^n u[-n-1] \rightarrow RoC := R_2 = |z| < \frac{1}{8}$$

Since the intersection of these RoCs are empty set, i.e. $R_1 \cap R_2 = \emptyset$, Z-transform of the given signal of $x[n]$ does not exist.

Example 2.4 Consider the following causal LTI system described as;

$$y[n] - \frac{3}{2}y[n-1] - y[n-2] = 7x[n] - \frac{3}{2}x[n-1]$$

(a) Derive the system function, $H(z)$
(b) Find the impulse response, $h[n]$ using inverse Z-transform

Solution 2.4

(a) Applying Z-transform both side of the given equation;

$$Y(z) - \frac{3}{2}z^{-1}Y(z) - z^{-2}Y(z) = 7X(z) - \frac{3}{2}z^{-1}X(z)$$

$$Y(z)\left(1 - \frac{3}{2}z^{-1} - z^{-2}\right) = X(z)\left(7 - \frac{3}{2}z^{-1}\right)$$

$$H(z) = \frac{Y(z)}{X(z)}$$

$$= \frac{7 - \frac{3}{2}z^{-1}}{1 - \frac{3}{2}z^{-1} - z^{-2}}$$

(b) In order to use Table 2.1, $H(z)$ should be rewritten using partial fractions technique;

$$H(z) = \frac{7 - \frac{3}{2}z^{-1}}{1 - \frac{3}{2}z^{-1} - z^{-2}}$$

$$= \frac{A}{1 + 2z^{-1}} + \frac{B}{1 - 2z^{-1}}$$

where A, and B are constants

After some algebraic calculations, it is obtained as $A = 5$, and $B = 2$

$$H(z) = \frac{5}{1 + 2z^{-1}} + \frac{2}{1 - 2z^{-1}}$$

RoC plays an important role while employing inverse Z-transform using Table 2.1. Since *RoC* is not specified in the example, there are there possibilities such that;

Case I: If $RoC := |z| > 2$

$$h[n] = \mathcal{Z}^{-1}\left\{\frac{5}{1+2z^{-1}} + \frac{2}{1-2z^{-1}}\right\}$$

$$= 2\left(-\frac{1}{2}\right)^n u[n] + 5\,(2)^n\, u[n]$$

Case II: If $RoC := \frac{1}{2} < |z| < 2$

$$h[n] = \mathcal{Z}^{-1}\left\{\frac{5}{1+2z^{-1}} + \frac{2}{1-2z^{-1}}\right\}$$

$$= 2\left(-\frac{1}{2}\right)^n u[n] - 5\,(2)^n\, u[-n-1]$$

Case III: If $RoC := \frac{1}{2} > |z|$

$$h[n] = \mathcal{Z}^{-1}\left\{\frac{5}{1+2z^{-1}} + \frac{2}{1-2z^{-1}}\right\}$$

$$= -2\left(-\frac{1}{2}\right)^n u[-n-1] - 5\,(2)^n\, u[-n-1]$$

2.3.1 Poles and Zeros of *H*(*z*)

In general, the system function $H(z)$ is expressed as ratio of two polynomial functions of z, such that;

$$H(z) = \frac{Q(z)}{P(z)}$$

The roots of $Q(z)$, i.e. where $Q(z) = 0$, are called the *zeros* of the system function, and the roots of $P(z)$, the values of z where $P(z) = 0$, are known as the*poles* of $H(z)$. On the *z-plane* the poles are represented by $\times$, whereas the zeros are shown as $\circ$.

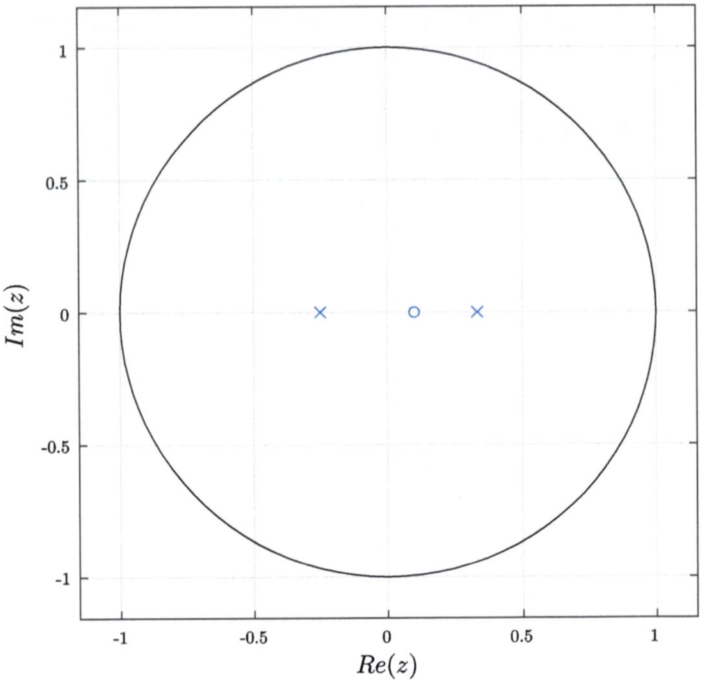

Fig. 2.3 The pole-zero diagram of Example 2.4

Example 2.5 Find the poles and zeros of the following system function.

$$H(z) = \frac{5 - \frac{1}{2}z^{-1}}{\left(1 - \frac{1}{3}z^{-1}\right)\left(1 + \frac{1}{4}z^{-1}\right)}.$$

Solution 2.5 The root of the polynomial located in the numerator part is $z = \frac{1}{10}$, which is *zero* of the system function. The roots for the denominator polynomial equation are $z = \frac{1}{3}$, and $z = -\frac{1}{4}$, which are called the poles of the system function. The corresponding pole-zero diagram is shown in Fig. 2.3.

2.3.2 *Analysis of LTI Systems Using System Function*

Since the system function $H(z)$ is a mathematical representation of the system in z-domain; it implicates important information about the system under investigation.

Causality

In time-domain representation of an LTI system, if the impulse response of the system $h[n]$ satisfies;

$$h[n] = 0 \;\; \forall \;\; n < 0$$

then such LTI system is called causal.

In z-domain, if the system is causal,then the RoC of the system function, $H(z)$, is the exterior of circle of radius outermost pole.

Stability

The RoC of the system function, $H(z)$, includes the unit circle, then the system is called stable.

Example 2.6 Consider an LTI system having the following system function;

$$H(z) = \frac{9 - \frac{11}{3}z^{-1}}{\left(1 + \frac{1}{3}z^{-1}\right)\left(1 - 3z^{-1}\right)}$$

(a) Find poles and zeros of the given system function.
(b) Derive the impulse response, $h[n]$ using inverse Z-transform if the system is stable and noncausal.
(c) Derive the impulse response, $h[n]$ using inverse Z-transform if the system is unstable and causal.
(d) Derive the impulse response, $h[n]$ using inverse Z-transform if the system is unstable and noncausal.

Solution 2.6

(a) The root of the numerator part is $z = \frac{11}{27}$, which is *zero* of the given system function. The roots for the denominator part are $z = -\frac{1}{3}$, and $z = 3$, which are called the poles of the system function. The corresponding pole-zero diagram is shown in Fig. 2.4.

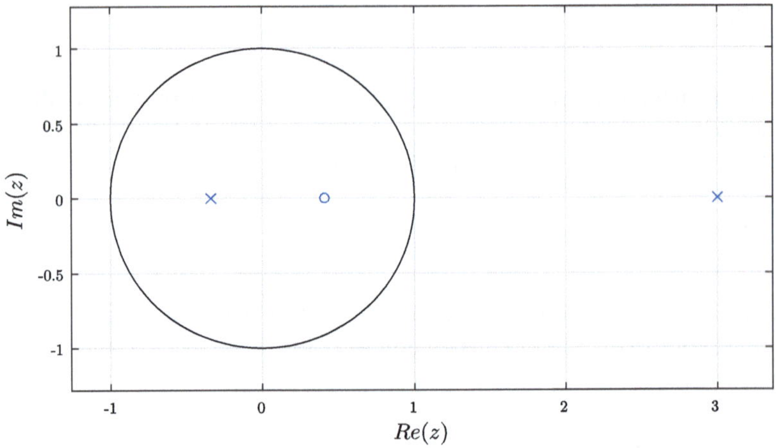

Fig. 2.4 The pole-zero diagram of Example 2.5

(b) It is convenient to rewrite $H(z)$ using partial fractions technique to use Table 2.1;

$$H(z) = \frac{9 - \frac{11}{3}z^{-1}}{\left(1 + \frac{1}{3}z^{-1}\right)\left(1 - 3z^{-1}\right)}$$

$$= \frac{A}{1 + \frac{1}{3}z^{-1}} + \frac{B}{1 - 3z^{-1}}$$

where A, and B are constants

After simple algebraic calculations, these constants are obtained as $A = 2$, and $B = 7$

$$H(z) = \frac{2}{1 + \frac{1}{3}z^{-1}} + \frac{7}{1 - 3z^{-1}}$$

If the system is stable and noncausal, then $RoC := \frac{1}{3} < |z| < 3$

$$h[n] = \mathcal{Z}^{-1}\left\{\frac{2}{1 + \frac{1}{3}z^{-1}} + \frac{7}{1 - 3z^{-1}}\right\}$$

$$= 2\left(-\frac{1}{3}\right)^n u[n] - 7\,(3)^n\, u[-n-1]$$

(c) If the system is unstable and causal, then $RoC := |z| > 3$

$$h[n] = \mathcal{Z}^{-1}\left\{\frac{2}{1+\frac{1}{3}z^{-1}} + \frac{7}{1-3z^{-1}}\right\}$$

$$= 2\left(-\frac{1}{3}\right)^n u[n] + 7\,(3)^n\,u[n]$$

(d) If the system is unstable and noncausal, then $RoC := |z| < \frac{1}{3}$

$$h[n] = \mathcal{Z}^{-1}\left\{\frac{2}{1+\frac{1}{3}z^{-1}} + \frac{7}{1-3z^{-1}}\right\}$$

$$= -2\left(-\frac{1}{3}\right)^n u[-n-1] - 7\,(3)^n\,u[-n-1]$$

Example 2.7 Consider an LTI system with

$$H(z) = \frac{1}{\left(1-2z^{-1}\right)\left(1-0.75z^{-1}\right)\left(1-0.25z^{-1}\right)}$$

where $RoC := 0.75 < |z| < 2$.

(a) Find the impulse response, $h(n)$
(b) Is the system causal?
(c) Is the system stable?

Solution 2.7

(a)

$$H(z) = \frac{1}{\left(1-2z^{-1}\right)\left(1-0.75z^{-1}\right)\left(1-0.25z^{-1}\right)}$$

$$= \frac{A}{1-2z^{-1}} + \frac{B}{1-0.75z^{-1}} + \frac{C}{1-0.25z^{-1}}$$

$$A = \left.\frac{1}{\left(1-0.75z^{-1}\right)\left(1-0.25z^{-1}\right)}\right|_{z=2} = \frac{1}{\left(\frac{5}{8}\right)\left(\frac{7}{8}\right)} = \frac{64}{35}$$

$$B = \left.\frac{1}{\left(1-2z^{-1}\right)\left(1-0.25z^{-1}\right)}\right|_{z=0.75} = -0.9$$

$$C = \left.\frac{1}{\left(1-2z^{-1}\right)\left(1-0.75z^{-1}\right)}\right|_{z=0.25} = \frac{1}{14}$$

$$H(z) = \underbrace{\frac{64/35}{1-2z^{-1}}}_{RoC:=|z|<2} - \underbrace{\frac{0.9}{1-0.75z^{-1}}}_{RoC:=|z|>0.75} + \underbrace{\frac{1/14}{1-0.25z^{-1}}}_{RoC:=|z|>0.25}$$

Since, the region of convergence is given as $RoC := 0.75 < |z| < 2$, the impulse response can be obtained using Table 2.1;

$$h[n] = \frac{-64}{35}(2)^n u[-n-1] - 0.9(0.75)^n u[n] + \frac{1}{14}(0.25)^n u[n]$$

(b) Since RoC is not given as "outside of the outermost pole" the system is not causal.
(c) The system is stable, because RoC include the unit circle, $|z| = 1$

Problems

2.1 Considering the following discrete time signal $x[n] = \left(\frac{1}{3}\right)^n u[n] + (4)^n u[n]$. It is;

(a) Right-sided
(b) Left-sided
(c) Two-sided
(d) None of them

2.2 For the following discrete time signal, $x[n] = \left(\frac{1}{7}\right)^n u[n] - (8)^n u[-n-1]$. The system has poles at;

(a) $\frac{1}{7}; -8$
(b) $\frac{-1}{7}; -8$
(c) $\frac{1}{7}; 8$
(d) $\frac{-1}{7}; 8$

2.3 Consider the following system $h[n] = -\left(\frac{1}{5}\right)^n u[-n-1] - (3)^n u[-n-1]$. The RoC of $H(z)$ is;

(a) $|z| < \frac{1}{5}$
(b) $|z| < 3$
(c) $\frac{1}{5} < |z| < 3$
(d) $5 < |z| < \frac{1}{3}$

2.4 Let a discrete time signal be $x[n] = \left(\frac{1}{7}\right)^n u[n] - (8)^n u[-n-1]$. The corresponding Z-transform is;

(a) $\frac{1}{1+\frac{1}{7}z^{-1}} - \frac{1}{1-8z^{-1}}$
(b) $\frac{1}{1-\frac{1}{7}z^{-1}} + \frac{1}{1-8z^{-1}}$

(c) $\frac{1}{1+\frac{1}{7}z^{-1}} + \frac{1}{1+8z^{-1}}$
(d) $\frac{1}{1-\frac{1}{7}z^{-1}} - \frac{1}{1+8z^{-1}}$

2.5 Find the inverse Z-transform of $X(z) = \frac{1}{1-4z^{-1}}$ RoC := $|z| > 4$

(a) $x[n] = (-4)^n u[-n-1]$
(b) $x[n] = \left(\frac{1}{4}\right)^n u[n]$
(c) $x[n] = (4)^n u[n]$
(d) $x[n] = \left(\frac{1}{4}\right)^n u[-n-1]$

2.6 Find the inverse Z-transform of

$$X(z) = 3z^2 + 5z + 7z^{-1} \quad RoC := 0 < |z| < \infty$$

(a) $x[n] = 3\delta[n-2] + 5\delta[n-1] + 7\delta[n+1]$
(b) $x[n] = 3\delta[n+2] + 5\delta[n+1] + 7\delta[n-1]$
(c) $x[n] = 3\delta[n+2] + 5\delta[n] + 7\delta[n+1]$
(d) $x[n] = 3\delta[n] + 5\delta[n-2] + 7\delta[n+1]$

2.7 For the unit impulse signal $\delta[n]$:

(a) The z-transform is equal to 1 and the region of convergence contains only the point $z = 1$
(b) The z-transform is equal to 1 and does not converge
(c) The z-transform is equal to 1 and the region of convergence contains all possible values of z
(d) The z-transform is equal to 1 and the region of convergence contains all possible values of z except for $z = 0$ and $z = \infty$

2.8 Calculate the Z-transform of the following signals, indicate the RoC for each case.

(a) $\left(\frac{1}{5}\right)^n u[n]$
(b) $-\left(\frac{1}{5}\right)^n u[-n-1]$
(c) $3\delta[n] + 2\delta[n-1]$
(d) $\left(\frac{1}{5}\right)^n (u[n-2] - u[n-4])$

2.9 Consider the following functions are given in z-domain; derive the all possible inverse z-transforms for each function using Tables 2.1 and 2.2. Discuss stability and causality for each case.

(a)

$$X(z) = \frac{1}{2 - \frac{1}{3}z^{-1}}$$

(b)

$$X(z) = \frac{2 - \frac{1}{4}z^{-1}}{1 - \frac{1}{3}z^{-1}}$$

(c)

$$X(z) = \frac{1 + 2z^{-1}}{1 + \frac{3}{10}z^{-1} - \frac{1}{10}z^{-2}}$$

(d)

$$X(z) = \frac{1 + z^{-1} + 2z^{-2}}{\left(1 + \frac{1}{2}z^{-1}\right)\left(1 - z^{-1}\right)\left(1 + \frac{1}{3}z^{-1}\right)}$$

2.10 A causal LTI system with the following system function

$$H(z) = \frac{1 - 2z^{-1}}{\left(1 + \frac{1}{2}z^{-1}\right)\left(1 + z^{-1}\right)}$$

Considering the following signal is applied as input to such system

$$x[n] = \left(\frac{1}{5}\right)^n u[n-1]$$

(a) Calculate the impulse response of the system, $h[n]$
(b) Derive $y[n]$ using convolution sum
(c) Calculate $y[n]$ using z-transform and inverse z-transform, then compare your finding with part (b)
(d) Discuss whether the system is stable or not

2.11 Plot pole-zero diagram for the following systems, indicate the RoC for each case.

(a)

$$H(z) = \frac{1}{\left(1 - \frac{2}{3}z^{-1}\right)\left(1 + 2z^{-1}\right)}; \quad \text{stable}$$

(b)

$$H(z) = \frac{1 - 2z^{-1}}{\left(1 - \frac{4}{5}z^{-1}\right)\left(1 - \frac{1}{2}z^{-1}\right)}; \quad \text{causal}$$

(c)

$$H(z) = \frac{1}{1 + 4z^{-1}}, \quad \text{unstable}$$

(d)

$$H(z) = \frac{1}{1 - 5z^{-1}}, \quad \text{noncausal}$$

2.12 Consider the following causal LTI system

$$y[n] - \frac{3}{8}y[n-1] - \frac{1}{32}y[n-2] = 4x[n] - \frac{1}{2}x[n-1]$$

(a) Derive the system function, $H(z)$
(b) Find the impulse response, $h[n]$ using inverse Z-transform

2.13 Consider an LTI system having the following impulse response

$$h[n] = 2^n u[n]$$

and the following signal is applied as input

$$x[n] = u[n] + \delta[n] + \delta[n-1]$$

(a) Derive $y[n]$ using convolution sum
(b) Calculate $y[n]$ using z-transform

2.14 If the input to an LTI system is

$$x[n] = \left(\frac{1}{2}\right)^n u[n] - (2)^n u[-n-1]$$

then, the corresponding output is obtained as;

$$y[n] = 6\left(\frac{1}{4}\right)^n u[n] + 3\left(\frac{1}{5}\right)^n u[n-1]$$

(a) Derive $H(z)$, and indicate the RoC
(b) Discuss whether the system is causal or not
(c) Is this system stable?

2.15 Find all possible impulse responses, $h[n]$, of the following system having the system function as;

$$H(z) = \frac{2 + \frac{11}{12}z^{-1} + \frac{7}{24}z^{-2}}{\left(1 + \frac{1}{2}z^{-1}\right)\left(1 - \frac{1}{3}z^{-1}\right)\left(1 + \frac{1}{4}z^{-1}\right)}$$

Discuss the stability of the system for each case.

2.16 Determine the Region of Convergence (RoC) for the Z-transform of the following signals:

(a) $x[n] = a^n u[n] + b^n u[-n-1]$, where $|a| < |b|$
(b) $x[n] = na^n u[n]$, where $|a| < 1$
(c) $x[n] = \delta[n-4] + \delta[n+2]$
(d) $x[n] = \begin{cases} a^n, & 0 \le n \le N-1 \\ 0, & \text{otherwise} \end{cases}$, where $|a| < \infty$

2.17 Given the Z-transform $X(z) = \frac{z}{z-0.5}$, determine the inverse Z-transform $x[n]$ for the following RoCs:

(a) $|z| > 0.5$
(b) $|z| < 0.5$

Discuss whether the system is causal and stable for each case.

2.18 A discrete-time LTI system is described by the difference equation:

$$y[n] - y[n-1] + 0.25y[n-2] = x[n] - 0.5x[n-1]$$

(a) Find the system function $H(z)$.
(b) Determine the impulse response $h[n]$ if the system is causal.
(c) Determine if the causal system is stable.

2.19 Find the Z-transform of the following signals:

(a) $x[n] = n(0.8)^n u[n]$
(b) $x[n] = \sin(\omega_0 n)u[n]$, where ω_0 is a constant
(c) $x[n] = 2^n u[-n] + (0.5)^n u[n]$
(d) $x[n] = u[n] - u[n-N]$, where N is a positive integer

2.20 Determine the inverse Z-transform of the following functions:

(a) $X(z) = \frac{z}{(z-1)(z-2)}$, $RoC : |z| > 2$
(b) $X(z) = \frac{z}{(z-1)(z-2)}$, $RoC : 1 < |z| < 2$
(c) $X(z) = \frac{z}{(z-1)(z-2)}$, $RoC : |z| < 1$
(d) $X(z) = \log(1 - az^{-1})$, $|z| > |a|$

2.21 Consider a causal LTI system with system function

$$H(z) = \frac{1 - z^{-1}}{1 - 0.8z^{-1}}$$

(a) Find the difference equation that relates the input $x[n]$ and output $y[n]$.
(b) Compute the output $y[n]$ for the input $x[n] = (0.5)^n u[n]$.

2.22 For the following system function, determine all possible regions of convergence (RoCs) and for each RoC, determine if the system is causal, stable, or both.

$$H(z) = \frac{z^2}{(z-0.2)(z-4)}$$

2.23 The input $x[n]$ and output $y[n]$ of a causal LTI system are related by the difference equation:

$$y[n] - \frac{1}{6}y[n-1] - \frac{1}{6}y[n-2] = x[n]$$

(a) Find the system function $H(z)$.
(b) Find the impulse response $h[n]$ of the system.
(c) Determine if the system is stable.

2.24 Determine the convolution of the following sequences using the Z-transform:

$$x[n] = u[n]$$

$$h[n] = (0.5)^n u[n]$$

2.25 A system has a transfer function

$$H(z) = \frac{1}{1 - 1.5z^{-1} + 0.5z^{-2}}$$

(a) Find all possible impulse responses for this system.
(b) For each impulse response, indicate whether the system is stable and/or causal.

2.26 If $X(z) = \frac{1}{1-az^{-1}}$ and $H(z) = \frac{1}{1-bz^{-1}}$, with RoCs $|z| > |a|$ and $|z| > |b|$ respectively, find the Z-transform of the output $Y(z)$ and the output sequence $y[n]$ when the input $x[n]$ is applied to the system with impulse response $h[n]$. Assume $|a| < 1$ and $|b| < 1$.

2.27 A discrete-time system has the following system function:

$$H(z) = \frac{z^2+1}{z^2-0.25}$$

(a) Determine all possible impulse responses for this system.
(b) For each case, specify whether the system is causal and stable.

Chapter 3
Linear Transformations: Discrete Time Fourier Transform (DTFT)

Fourier transforms are very capable tools especially in filtering applications aiming to observe the behaviour of the system under various frequencies. In theory, any necessary calculations can be done in time domain. However, aforementioned transforms can dramatically reduce not only the computational cost but also the time spent for calculations. Therefore, they are inevitably employed in signal processing applications. In this chapter, Discrete time Fourier transform (DTFT) is introduced in detail.

Considering that $x[n]$ is a discrete signal, then Discrete Time Fourier Transform (DTFT) of $x[n]$ can be calculated as;

$$X(e^{j\omega}) = \sum_{n=-\infty}^{\infty} x[n]\mathrm{e}^{-j\omega n} \quad \text{Analysis equation} \tag{3.1}$$

and it is notated as

$$X(e^{j\omega}) = \mathfrak{F}\{x[n]\}$$

Under the condition that $X(e^{j\omega})$ is given, then $x[n]$ can be obtained using Inverse DTFT, such as;

$$x[n] = \mathfrak{F}^{-1}\left\{X(e^{j\omega})\right\} = \frac{1}{2\pi}\int_{-\pi}^{\pi} X(e^{j\omega})e^{j\omega n}\mathrm{d}\omega \quad \text{Synthesis equation} \tag{3.2}$$

M. Catak and T. Allahviranloo, *Discrete Random Processes*, Studies in Systems, Decision and Control 644, https://doi.org/10.1007/978-3-032-08630-3_3

In general, $x[n]$ and $X(e^{j\omega})$ are called DTFT pair, then the relationship is shown as;

$$x[n] \leftrightarrow X(e^{j\omega})$$

- $X(e^{j\omega})$ and $x[n]$ functions are not symmetrical. $X(e^{j\omega})$ can be defined as summation of multiplications of $x[n]$ and complex exponentials at specific frequency values over infinite terms. On the other hand, $x[n]$ is derived as integration over a finite interval of 2π (typically from $-\pi$ to π).
- The signal, $x[n]$ is discrete over time, and it is aperiodic.
- However, $X(e^{j\omega})$ is a continuous function in frequency (ω) domain, and it is periodic with the fundamental period of 2π.

Example 3.1 If a discrete time signal is given as;

$$x[n] = (\alpha)^n u[n]$$

Calculate $X(e^{j\omega})$, specify the condition where $X(e^{j\omega})$ exists.

Solution 3.1

DTFT of $x[n]$ is calculated as;

$$\begin{aligned} X\left(e^{j\omega}\right) &= \sum_{n=-\infty}^{\infty} (\alpha)^n u[n] e^{-j\omega n} \\ &= \sum_{n=0}^{\infty} \left(\alpha e^{-j\omega}\right)^n \end{aligned}$$

Using the geometric summation formula

$$\sum_{n=0}^{\infty} r^n = \frac{1}{1-r}, \text{ where } \quad |r| < 1$$

then,

$$X\left(e^{j\omega}\right) = \frac{1}{1 - ae^{-j\omega}}, \text{ where } \quad |\alpha| < 1$$

Example 3.2 Consider the following constant sequence of discrete time signal;

$$x[n] = 1$$

Calculate $X(e^{j\omega})$, specify the condition where $X(e^{j\omega})$ exists.

Solution 3.2

Let $x[n] = 1$, for all n, whose DTFT $X(e^{j\omega})$ can be derived by means of the sum of the complex exponentials;

$$X(e^{j\omega}) = \sum_{n=-\infty}^{\infty} 1e^{-j\omega n}$$

$$= 2\pi \sum_{m=-\infty}^{\infty} \delta(\omega - 2\pi m)$$

Therefore, the DTFT pair can be written as;

$$x[n] = 1 \leftrightarrow X(e^{j\omega}) = 2\pi \sum_{m=-\infty}^{\infty} \delta(\omega - 2\pi m)$$

3.1 Properties of DTFT

In this section some important properties of DTFT will be discussed.

3.1.1 *Periodicity*

DTFT of a signal $x[n]$ is periodic in frequency domain, with the fundamental period of 2π.

$$X(e^{j\omega}) = X(e^{j(\omega+2\pi)})$$

The reason for the periodicity lays on discrete time complex exponential is intrinsically periodic since it is literally linear combination of sinusoidal functions. Therefore, the domain of $X(e^{j\omega})$ is stated over any interval of 2π. Accordingly, when we need to classify the frequency as *high* and *low*, a meticulous action is necessary. *Low*

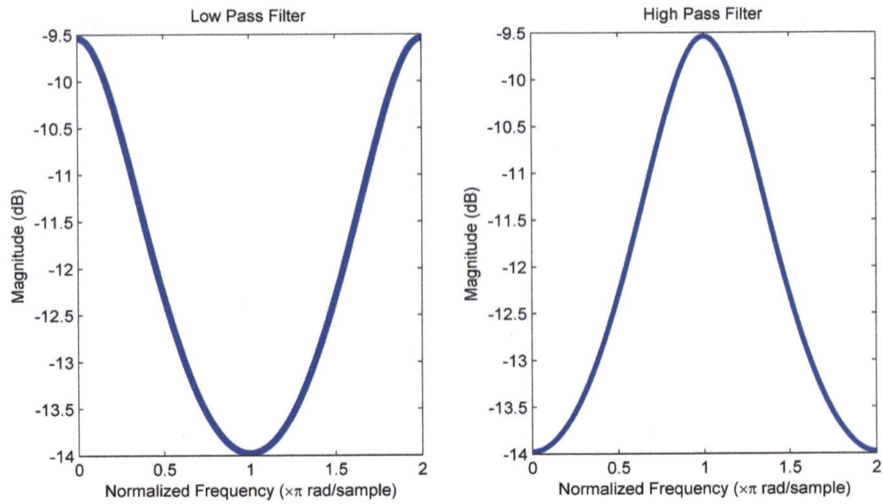

Fig. 3.1 An illustrative example for low-pass and high-pass filters

frequency part of a discrete signal refer to the around even multiplies of π, while *high* frequency part of a discrete signal near to the odd multiplies of π. An illustrative example for low pass filter and high pass filter is depicted in Fig. 3.1.

3.1.2 Linearity

Since summation operator is basically a linear operator, referring to Analysis equation, DTFT is a linear transformations. Therefore, for any constant value of α;

(i) $\mathfrak{F}\{\alpha x[n]\} = \alpha X(e^{j\omega})$
(ii) $\mathfrak{F}\{x_1[n] + x_2[n]\} = \mathfrak{F}\{x_1[n]\} + \mathfrak{F}\{x_2[n]\}$

In general it can be written as;

$$X(e^{j\omega}) = \mathfrak{F}\left\{\sum_{i=1}^{\infty} \alpha_i x_i[n]\right\} = \sum_{i=1}^{\infty} \alpha_i \mathfrak{F}\{x_i[n]\} = \sum_{i=1}^{\infty} \alpha_i X_i(e^{j\omega})$$

3.1.3 Time Shift

If it is known that

$$x[n] \leftrightarrow X(e^{j\omega})$$

Considering a time shift n_0 in $x[n]$ yields;

$$\begin{aligned}
\mathfrak{F}\{x[n-n_0]\} &= \sum_{n=-\infty}^{\infty} x[n-n_0]e^{-j\omega n} \\
&\text{by changing variable } k = n - n_0 \\
&= \sum_{k=-\infty}^{\infty} x[k]e^{-j\omega(n_0+k)} \\
&= \sum_{k=-\infty}^{\infty} e^{-j\omega n_0} x[k]e^{-j\omega(k)} \\
&= e^{-j\omega n_0} \sum_{k=-\infty}^{\infty} x[k]e^{-j\omega(k)}
\end{aligned}$$

hence, it can be written as;

$$x[n-n_0] \leftrightarrow e^{-j\omega n_0} X(e^{j\omega}) \tag{3.3}$$

3.1.4 Frequency Shift

If it is given that

$$x[n] \leftrightarrow X(e^{j\omega})$$

Considering a frequency shift of ω_0, one has

$$\mathfrak{F}^{-1}\left\{X(e^{j(\omega-\omega_0)})\right\} = \frac{1}{2\pi}\int_{-\pi}^{\pi} X(e^{j(\omega-\omega_0)})e^{j\omega n}\mathrm{d}\omega$$

by changing variable $\xi = \omega - \omega_0$, and $\mathrm{d}\xi = \mathrm{d}\omega$

$$= \frac{1}{2\pi}\int_{-\pi}^{\pi} X(e^{j\xi})e^{j(\xi+\omega_0)n}\mathrm{d}\xi$$

$$= e^{j\omega_0 n}\frac{1}{2\pi}\int_{-\pi}^{\pi} X(e^{j\xi})e^{j\xi n}\mathrm{d}\xi$$

then, it can be expressed as;

$$e^{j\omega_0 n}x[n] = \mathfrak{F}^{-1}\left\{X(e^{j(\omega-\omega_0)})\right\}$$

$$e^{j\omega_0 n}x[n] \leftrightarrow X(e^{j(\omega-\omega_0)}) \tag{3.4}$$

3.1.5 Differentiation in Frequency Domain

Let

$$X(e^{j\omega}) = \sum_{n=-\infty}^{\infty} x[n]e^{-j\omega n} \qquad \text{Analysis equation}$$

Assuming that $X(e^{j\omega})$ is a continuous and differentiable function, if one take derivative of both sides with respect to ω,

$$\frac{\mathrm{d}X\left(e^{j\omega}\right)}{\mathrm{d}\omega} = \sum_{n=-\infty}^{+\infty} -jnx[n]e^{-j\omega n}$$

Since, DTFT is a linear operator, if both sides are multiplied with j, then

$$j\frac{dX\left(e^{j\omega}\right)}{\mathrm{d}\omega} = \sum_{n=-\infty}^{+\infty} nx[n]e^{-j\omega n}$$

which is DTFT of $nx[n]$, therefore

$$j\frac{\mathrm{d}X(e^{j\omega})}{\mathrm{d}\omega} \leftrightarrow nx[n] \tag{3.5}$$

3.1.6 Convolution Sum in Time Domain

Assuming that

$$x_1[n] \leftrightarrow X_1(e^{j\omega})$$

$$x_2[n] \leftrightarrow X_2(e^{j\omega})$$

and $y[n]$ is defined as the convolution sum sum of $x_1[n]$ and $x_2[n]$;

$$\begin{aligned} y[n] &= x_1[n] * x_2[n] \\ &= \sum_{k=-\infty}^{\infty} x_1[k]x_2[n-k] \end{aligned}$$

$$\begin{aligned} Y(e^{j\omega}) &= \sum_{n=-\infty}^{\infty} Y[n]e^{-j\omega n} \\ &= \sum_{n=-\infty}^{\infty}\sum_{k=-\infty}^{\infty} x_1(k)x_2(n-k)e^{-j\omega n} \\ &= \sum_{k=-\infty}^{\infty} x_1(k) \underbrace{\sum_{n=-\infty}^{\infty} x_2(n-k)e^{-j\omega n}}_{e^{-j\omega k}X_2(e^{j\omega})} \\ &= \left(\sum_{k=-\infty}^{\infty} x_1(k)e^{-j\omega k}\right) X_2(e^{j\omega}) \\ &= X_1(e^{j\omega})X_2(e^{j\omega}) \end{aligned}$$

then;

$$Y(e^{j\omega}) = X_1(e^{j\omega})X_2(e^{j\omega}) \tag{3.6}$$

3.1.7 Multiplication in Time Domain or Convolution in Frequency Domain

Let $y[n] = x_1[n]x_2[n]$, and then assuming the following transformations exist,

$$x_1[n] \leftrightarrow X_1(e^{j\omega})$$

$$x_2[n] \leftrightarrow X_2(e^{j\omega})$$

$$y[n] \leftrightarrow Y(e^{j\omega})$$

$$\begin{aligned} Y\left(e^{j\omega}\right) &= \sum_{n=-\infty}^{\infty} y[n]e^{-j\omega n} \\ &= \sum_{n=-\infty}^{\infty} x_1[n]x_2[n]e^{-j\omega n} \end{aligned}$$

$x_1[n]$ can be defined as;

$$x_1[n] = \frac{1}{2\pi}\int_{-\pi}^{\pi} X_1\left(e^{j\xi}\right)e^{j\xi n}\mathrm{d}\xi,$$

then,

$$Y\left(e^{j\omega}\right) = \sum_{n=-\infty}^{\infty} x_2[n]\left\{\frac{1}{2\pi}\int_{-\pi}^{\pi} X_1\left(e^{j\xi}\right)e^{j\xi n}\mathrm{d}\xi\right\}e^{-j\omega n}.$$

if the order of summation and integration is shifted,

$$Y\left(e^{j\omega}\right) = \frac{1}{2\pi}\int_{-\pi}^{\pi} X_1\left(e^{j\xi}\right)\underbrace{\sum_{n=-\infty}^{\infty} x_2[n]e^{-j(\omega-\xi)n}\,\mathrm{d}\xi}_{X_2\left(e^{j(\omega-\xi)}\right)}$$

Therefore, it can be stated as;

$$Y\left(e^{j\omega}\right) = \frac{1}{2\pi}\int_{-\pi}^{\pi} X_1\left(e^{j\xi}\right)X_2\left(e^{j(\omega-\xi)}\right)\mathrm{d}\xi$$

$$x_1[n]x_2[n] \leftrightarrow \frac{1}{2\pi}\int_{-\pi}^{\pi} X_1(e^{j\xi})X_2(e^{j(\omega-\xi)})\mathrm{d}\xi \tag{3.7}$$

3.1.8 Modulation

Modulation has significant applications in communication area. Let $x[n]$ contains information data needed to be transmitted, instead of sending it as itself, it is more convenient put the information into an "envelope", then transmit it. Here, $\cos(\omega_0 n)$ is assumed an envelope.

$$x[n]\cos(\omega_0 n) = x[n]\frac{1}{2}\left(e^{j\omega_0 n} + e^{-j\omega_0 n}\right)$$

Using the frequency shift property;

$$\begin{aligned} &x[n]\cos(\omega_0 n) \\ &= x[n]\frac{1}{2}\left(e^{j\omega_0 n} + e^{-j\omega_0 n}\right) \quad \leftrightarrow \pi \sum_{m=-\infty}^{\infty} X\left(e^{j(\omega-\omega_0-2\pi m)}\right) + X\left(e^{j(\omega+\omega_0-2\pi m)}\right) \end{aligned}$$

$$x[n]\cos(\omega_0 n) \leftrightarrow \pi \sum_{m=-\infty}^{\infty} X\left(e^{j(\omega-\omega_0-2\pi m)}\right) + X\left(e^{j(\omega+\omega_0-2\pi m)}\right) \tag{3.8}$$

Example 3.3 If $x[n] = \cos(\omega_0 n)$, derive $X(e^{j\omega})$

Solution 3.3

$\cos(\omega_0 n)$ can be expressed using complex exponentials as:

$$x[n] = \cos(\omega_0 n) = \frac{1}{2}\left(e^{j\omega_0 n} + e^{-j\omega_0 n}\right)$$

By employing the frequency shift property;

$$x[n] \leftrightarrow X(e^{j\omega}) \Longrightarrow x[n]e^{j\omega_0 n} \leftrightarrow X\left((e^{j\omega})\right)$$

$$1 \leftrightarrow 2\pi\delta(\omega) \Longrightarrow 1 \cdot e^{j\omega_0 n} \leftrightarrow \pi \sum_{m=-\infty}^{\infty} \delta(\omega - \omega_0 - 2\pi m)$$

3.2 Some Commonly Used DTFT Pairs

Some commonly used DTFT pairs are shown in Table 3.1.

Some important properties of DTFT are depicted in Table 3.2.

3.3 Autocorrelation, $c_{xx}[n]$

Autocorrelation function is widely used in statistical signal processing area, especially in estimation-detection applications. Considering a real valued sequence $x[n]$, the autocorrelation of $x[n]$ can be derived as;

$$c_{xx}[n] = \sum_{m=-\infty}^{\infty} x[k]x[n+k]$$

which can be explain using the well-known convolution sum, such that;

$$c_{xx}[n] = x[n] * x[-n] = \sum_{m=-\infty}^{\infty} x[k]x[-(-n)+k]$$

The variable n in $c_{xx}[n]$ is called *lag*, and $c_{xx}[n]$ gets its maximum at $n = 0$, which equals to the energy of the sequence.

$$E_x = c_{xx}[0]$$

Table 3.1 Some commonly used DTFT pairs

$x[n]$	$X\left(e^{j\omega}\right)$
$\delta[n]$	1
$\delta[n-n_0]$	$e^{-j\omega n_0}$
$\alpha^n u[n],\quad \lvert\alpha\rvert < 1$	$\frac{1}{1-\alpha e^{-j\omega}}$
$u[n]$	$\frac{1}{1-e^{-j\omega}} + \pi \sum_{k=-\infty}^{\infty} \delta(\omega + 2\pi k)$
$\begin{cases} 1, 0 \le n \le M \\ 0, \text{ otherwise} \end{cases}$	$\frac{\sin\left[\omega\left(\frac{M+1}{2}\right)\right]}{\sin\left(\frac{\omega}{2}\right)} e^{-\frac{j\omega M}{2}}$
$\frac{\sin(\omega_c n)}{\pi n}, 0 < \omega_c \le \pi$	$\begin{cases} 1, \ \lvert\omega\rvert < \omega_c \\ 0, \ \omega_c < \lvert\omega\rvert \le \pi \end{cases}$
$= (n+1)\alpha^n u[n]$	$\frac{1}{(1-\alpha e^{-j\omega})^2}$

Table 3.2 Some commonly used properties of DTFT

Property	Sequence	Transform
	$x[n]$	$X\left(e^{j\omega}\right)$
	$x_1[n]$	$X_1\left(e^{j\omega}\right)$
	$x_2[n]$	$X_2\left(e^{j\omega}\right)$
Periodicity	$X(e^{j\omega}) = X(e^{j(\omega+2\pi)})$	
Linearity	$\mathfrak{F}\left\{\sum_{i=1}^{\infty} \alpha_i x_i[n]\right\} = \sum_{i=1}^{\infty} \alpha_i \mathfrak{F}\{x_i[n]\} = \sum_{i=1}^{\infty} \alpha_i X_i(e^{j\omega})$	
Time-shift	$x[n-n_0] \leftrightarrow e^{-j\omega n_0} X(e^{j\omega})$	
Frequency shift	$e^{j\omega_0 n} x[n] \leftrightarrow X(e^{j(\omega-\omega_0)})$	
Differentiation in ω	$j\frac{dX(e^{j\omega})}{d\omega} = \sum_{n=-\infty}^{\infty} n x[n] e^{-j\omega n}$	
Convolution in time	$y[n] = x_1[n] * x_2[n] \leftrightarrow Y(e^{j\omega}) = X_1(e^{j\omega}) X_2(e^{j\omega})$	
Convolution in ω	$x_1[n] x_2[n] \leftrightarrow \frac{1}{2\pi} \int_{-\pi}^{\pi} X_1(e^{j\xi}) X_2(e^{j(\omega-\xi)}) d\xi$	
Modulation	$x[n]\cos(\omega_0 n) \leftrightarrow \pi \sum_{m=-\infty}^{\infty} X\left(e^{j(\omega-\omega_0-2\pi m)}\right) + X\left(e^{j(\omega+\omega_0-2\pi m)}\right)$	

By applying DTFT to autocorrelation function with using the convolution property;

$$C_{xx}(e^{j\omega}) = X(e^{j\omega})X(e^{-j\omega}) \quad = \left|X(e^{j\omega})\right|^2 \tag{3.9}$$

Thereafter, the energy of a discrete time signal can be calculated as;

$$\begin{aligned} E_x = c_{xx}(0) &= \sum_{m=-\infty}^{\infty} |x[n]|^2 \\ &= \frac{1}{2\pi} \int_{-\pi}^{\pi} C_{xx}(e^{j\omega}) d\omega \end{aligned} \tag{3.10}$$

3.4 LTI System Analysis Using DTFT

Convolution sum operator defines the relationship between the input, $x[n]$, and the corresponding output, $y[n]$, in LTI systems, such as

$$y[n] = x[n] * h[n] = h[n] * x[n]$$

where $h[n]$ is the impulse response, that is the response of the system to the unit impulse function $\delta[n]$.

Consider on LTI system with the input $x[n] = e^{j\omega n}$, the output can be derived as

$$\begin{aligned} y[n] &= \sum_{k=\infty}^{\infty} h[k]x[n-k] \\ &= \sum_{k=-\infty}^{\infty} h[k]e^{j\omega n}e^{-j\omega k} \\ &= \sum_{k=-\infty}^{\infty} e^{j\omega n}h[k]e^{-j\omega k} \\ &= e^{j\omega n} \underbrace{\sum_{k=-\infty}^{\infty} h[k]e^{-j\omega k}}_{H\left(e^{j\omega}\right)} \\ &= e^{j\omega n} H\left(e^{j\omega}\right) \end{aligned} \tag{3.11}$$

Here, $e^{j\omega n}$ is called *eigenfunction* of the system, and $H\left(e^{j\omega}\right)$ the *eigenvalue* of the system. In general. $H\left(e^{j\omega}\right)$ is known as the *Frequency Response* of the system since it describes the response of the system to a specific frequency defined in a complex exponential.

Example 3.3: Sinusoidal Input

Let the input of an LTI System be

$$x[n] = A\cos\left(\omega_0 n + \theta\right)$$

Calculate the corresponding output interns of frequent response.

Solution 3.3

$$\begin{aligned} X[n] &= A\cos(\omega_0 n + \theta) \\ &= A\frac{1}{2}\left[e^{j(\omega_0 n+\theta)} + e^{-j(\omega_0 n+\theta)}\right] \\ &= \frac{A}{2}e^{j\theta}e^{j\omega_0 n} + \frac{A}{2}e^{-j\theta}e^{-j\omega_0 n} \end{aligned}$$

The corresponding output will be defined as:

$$y[n] = \frac{A}{2}e^{j\theta}H\left(e^{j\omega_0}\right)e^{j\omega_0 n} + \frac{A}{2}e^{-j\theta}H\left(e^{-j\omega_0}\right)e^{-j\omega_0 n}$$

Under the condition of $h[n]$ is a real-valued sequence, then

$$\begin{aligned} H\left(e^{-J\omega_0}\right) &= H^*\left(e^{j\omega_0}\right); \text{ (* means complex conjugate)} \\ y[n] &= A\left|H\left(e^{j\omega_0}\right)\right|\cos(\omega_0 n + \theta + \phi) \end{aligned}$$

where $\phi = \angle H\left(e^{j\omega_0}\right)$, phase angle of the frequency response.

Example 3.4 Consider the following system, the input-output relation expressed by means of linear constant coefficient difference equation

$$y[n] = \frac{1}{3}y[n-1] + x[n] - \frac{1}{5}x[n-1]$$

(a) Find the frequent response, $H\left(e^{j\omega}\right)$ of the system
(b) Derive the impulse response, $h[n]$

Solution 3.4

(a) By definition, if the input is assigned as $\delta[n]$, then the corresponding output will be $h[n]$. Therefore,

$$h[n] = \frac{1}{3}h[n-1] + \delta[n] - \frac{1}{5}\delta[n-1]$$

By applying DTFT on both sides (since DTFT is a linear operator)

$$H\left(e^{j\omega}\right) = \frac{1}{3}e^{-j\omega}H\left(e^{j\omega}\right) + 1 - \frac{1}{5}e^{-j\omega}$$

$$H\left(e^{j\omega}\right)\left[1 - \frac{1}{3}e^{-j\omega}\right] = 1 - \frac{1}{5}e^{-j\omega}$$

$$H\left(e^{j\omega}\right) = \frac{1 - \frac{1}{5}e^{-j\omega}}{1 - \frac{1}{3}e^{-j\omega}}$$

$$= \frac{1}{1 - \frac{1}{3}e^{-j\omega}} - \frac{\frac{1}{5}e^{-j\omega}}{1 - \frac{1}{3}e^{-j\omega}}$$

(b) Using Tables 3.1 and 3.2

$$h[n] = \mathfrak{F}^{-1}\left\{H\left(e^{j\omega}\right)\right\}$$

$$= \mathfrak{F}^{-1}\left\{\frac{1}{1 - \frac{1}{3}e^{-j\omega}}\right\} - \mathfrak{F}^{-1}\left\{\frac{1}{5}\frac{e^{-j\omega}}{1 - \frac{1}{3}e^{-j\omega}}\right\}$$

$$= \left(\frac{1}{3}\right)^{n} u(n) - \frac{1}{5}\left(\frac{1}{3}\right)^{n-1} u(n-1).$$

Problems

3.1 Calculate the DTFT of the following signals.

(a) $\left(\frac{1}{2}\right)^n u[n]$
(b) $-\left(\frac{1}{3}\right)^n u[n-1]$
(c) $4\delta[n] - 3\delta[n-2]$
(d) $\left(\frac{1}{2}\right)^n (u[n+2] - u[n-7])$

3.2 Find the DTFT of each of the following signals:

(a) $x[n] = 2\delta[n-3] + 3\delta[n-4]$
(b) $x[n] = \delta[n-1] + 4\delta[n-2] - \delta[n-3]$
(c) $x[n] = 4u[n-2] - 4u[n-8]$
(d) $x[n] = \frac{\sin(\frac{2}{3}\pi n)}{6\pi n}$

3.3 Consider an LTI system defined by the following difference equation

$$y[n] = 3y[n-1] + 3x[n] + 3x[n-1] + 3x[n-2]$$

(a) Find the frequent response, $H\left(e^{j\omega}\right)$ of the system
(b) Derive the impulse response, $h[n]$

3.4 Assume that $x[n]$ and $y[n]$ are both discrete time aperiodic signals and that the DTFT of $x[n]$ is $X\left(e^{j\omega}\right)$. What is the DTFT of $y[n] = e^{j2\omega_0 n}x[n]$?

(a) $Y\left(e^{j\omega}\right) = X\left(e^{j(\omega-\omega_0)}\right)$
(b) $Y\left(e^{j\omega}\right) = X\left(e^{j(\omega+2\omega_0)}\right)$
(c) $Y\left(e^{j\omega}\right) = X\left(e^{j(\omega-2\omega_0)}\right)$
(d) $Y\left(e^{j\omega}\right) = e^{j2w_0 n}X\left(e^{j\omega}\right)$

3.5 Which of the following statements is correct for the DTFT of any signal?

(a) Continuous and Periodic with a period of 2π
(b) Discrete and aperiodic
(c) Non-periodic and continuous
(d) Discrete and periodic

3.6 The DTFT of the following discrete-time signal: $x[n] = (0.3)^n u[n]$

(a) $X\left(e^{j\omega}\right) = 2\cos(0.3\omega) + 1$
(b) $X\left(e^{j\omega}\right) = 0.3 - \cos(0.3\omega)$
(c) $X\left(e^{j\omega}\right) = \frac{1}{(1-0.3e^{-j\omega})}$
(d) $X\left(e^{j\omega}\right) = \frac{0.3}{(1+e^{j\omega})}$

3.7 Assume that $x[n]$ and $y[n]$ are both discrete time aperiodic signals and that the DTFT of $x[n]$ is $X\left(e^{j\omega}\right)$. What is the DTFT of $y[n] = x[n] - 2x\left[n - n_0\right]$?

(a) $Y\left(e^{j\omega}\right) = X\left(e^{j\omega}\right) - 2X\left(e^{j(\omega+2\omega_0)}\right)$
(b) $Y\left(e^{j\omega}\right) = X\left(e^{j\omega}\right) - 2X\left(e^{j\omega}\right)e^{-j\omega n_0}$
(c) $Y\left(e^{j\omega}\right) = X\left(e^{j\omega}\right) + 2X\left(e^{j\omega}\right)e^{j\omega_0 n}$
(d) $Y\left(e^{j\omega}\right) = X\left(e^{j\omega}\right) - 2X\left(e^{j(\omega-\omega_0)}\right)$

3.8 Given that $x[n]$ and $y[n]$ are both discrete time aperiodic signals, and that the DTFT of $x[n]$ is $X\left(e^{j\omega}\right)$. What is the DTFT of $y[n] = 2x[n] + nx[n]$?

(a) $Y\left(e^{j\omega}\right) = 2X\left(e^{j\omega}\right) - 2X\left(e^{j\omega}\right)e^{-j\omega n_0}$
(b) $Y\left(e^{j\omega}\right) = 2X\left(e^{j\omega}\right) + \omega X\left(e^{j\omega}\right)$
(c) $Y\left(e^{j\omega}\right) = 2X\left(e^{j\omega}\right) + 2X\left(e^{j\omega}\right)e^{j\omega n_0}$
(d) $Y\left(e^{j\omega}\right) = 2X\left(e^{j\omega}\right) + j\frac{dX\left(e^{j\omega}\right)}{d\omega}$

3.9 Calculate the inverse DTFT of the following $X(e^{j\omega})$:

(a) $X(e^{j\omega}) = 2\pi\delta(\omega),\ -\pi < \omega \leq \pi$
(b) $X(e^{j\omega}) = e^{-j4\omega}$
(c) $X(e^{j\omega}) = \begin{cases} 1, & |\omega| \leq \omega_c \\ 0, & \omega_c < |\omega| \leq \pi \end{cases}$, where $0 < \omega_c < \pi$
(d) $X(e^{j\omega}) = 2 + 4\cos(\omega) + 6\cos(2\omega)$

3.10 Determine the DTFT of the following signals:

(a) $x[n] = a^{|n|}$, where $|a| < 1$
(b) $x[n] = \begin{cases} n+1, & 0 \le n \le N-1 \\ 0, & \text{otherwise} \end{cases}$, where N is a positive integer
(c) $x[n] = \delta[n] - \delta[n-1]$
(d) $x[n] = \sum_{k=-\infty}^{\infty} \delta[n - kM]$, where M is a positive integer

3.11 Consider an LTI system with frequency response

$$H(e^{j\omega}) = \frac{1}{1 - 0.5e^{-j\omega}}$$

(a) Find the impulse response $h[n]$ of the system.
(b) Determine the output $y[n]$ of the system when the input is $x[n] = \delta[n] - 0.5\delta[n-1]$.

3.12 Let $x[n]$ have DTFT $X(e^{j\omega})$. Find the DTFT of the following signals in terms of $X(e^{j\omega})$:

(a) $y[n] = x[-n]$
(b) $y[n] = x[n] * x[n]$ (convolution)
(c) $y[n] = x[n]e^{j\omega_0 n}$, where ω_0 is a constant
(d) $y[n] = nx[n]$

3.13 A discrete-time LTI system has the impulse response

$$h[n] = \begin{cases} 1, & 0 \le n \le 2 \\ 0, & \text{otherwise} \end{cases}$$

(a) Find the frequency response $H(e^{j\omega})$ of the system.
(b) Plot the magnitude and phase response of the system.

3.14 Determine the output $y[n]$ of an LTI system with frequency response

$$H(e^{j\omega}) = \frac{1}{1 - 0.8e^{-j\omega}}$$

when the input is

$$x[n] = \cos(\omega_0 n)u[n]$$

3.15 Find the inverse DTFT of

$$X(e^{j\omega}) = \begin{cases} e^{-j2\omega}, & |\omega| < \frac{\pi}{4} \\ 0, & \frac{\pi}{4} \le |\omega| \le \pi \end{cases}$$

3.16 If the DTFT of $x[n]$ is $X(e^{j\omega})$, find the DTFT of $y[n] = x[2n]$.

3.17 A causal LTI system is described by the difference equation

$$y[n] - 0.5y[n-1] = x[n] + 0.25x[n-1]$$

(a) Find the frequency response $H(e^{j\omega})$ of the system.
(b) Find the impulse response $h[n]$ of the system.

3.18 Compute the DTFT of the following finite-duration signal:

$$x[n] = \begin{cases} 1, & -2 \leq n \leq 2 \\ 0, & \text{otherwise} \end{cases}$$

and sketch its magnitude spectrum.

3.19 An LTI system has the frequency response

$$H(e^{j\omega}) = e^{-j\omega}(1 + e^{-j\omega})$$

Determine the output $y[n]$ when the input is

$$x[n] = \delta[n] + 2\delta[n-1]$$

3.20 If $x[n] \overset{DTFT}{\longleftrightarrow} X(e^{j\omega})$, find the inverse DTFT of $\frac{dX(e^{j\omega})}{d\omega}$.

Chapter 4
Linear Transformations: Discrete Fourier Transform (DFT) and Its Implementation: Fast Fourier Transform (FFT)

4.1 Discrete Fourier Transform (DFT)

A discrete time system can be investigated by using Z transform, for example to show some system features like stability and causality. When $|z|$ is equal to unity, or $z = e^{j\omega}$, the Z transform and DTFT are equivalent. Since ω, the angular frequency, is the unique variable parameter for $z = e^{j\omega}$, DTFT is utilized to examine and understanding a discrete time system in the frequency domain. The associated DTFT for any appropriate impulse response, $h[n]$, has a fundamental period in the frequency domain of 2π. It is a continuous and periodic function. As a result, DTFT is inadequate for system analysis implementations in the realm of digital signal processing. However with analytical calculations taken into consideration, it retains its significance.

At this point, Discrete Fourier Transform (DFT), an alternative Fourier transform of discrete signals, may have a significant impact on digital signal processing applications.

Consider a finite-length discrete signal, which is defined only where $0 \leq n \leq N-1$, otherwise, $x[n] = 0$. The corresponding DFT of $x[n]$ can be calculated as

$$X[k] = \sum_{n=0}^{N-1} x[n] e^{-j\frac{2\pi}{N}kn} \quad \text{Analysis Equation} \tag{4.1}$$

Similarly the finite-length sequence $x[n]$ can be obtained from its DFT by means of applying inverse-DFT, such that;

M. Catak and T. Allahviranloo, *Discrete Random Processes*, Studies in Systems, Decision and Control 644, https://doi.org/10.1007/978-3-032-08630-3_4

$$x[n] = \frac{1}{N}\sum_{k=0}^{N-1} X[k]e^{-j\frac{2\pi}{N}kn} \quad \text{Synthesis Equation} \tag{4.2}$$

For the sake of simplicity, it is defined as;

$$W_N = e^{-j\frac{2\pi}{N}}$$

then

$$X[k] = \sum_{n=0}^{N-1} x[n]W_N^{kn}$$

and

$$x[n] = \frac{1}{N}\sum_{k=0}^{N-1} X[k]W_N^{kn}$$

Please note that, $X[k] = 0$ if k outside the interval of $0 \leq k \leq N-1$; and $x[n] = 0$ while n is not in the interval of $0 \leq n \leq N-1$. The relationship between $x[n]$ and $X[k]$ is notated as

$$x[n] \overset{\mathfrak{DFT}}{\longleftrightarrow} X[k]$$

Example 4.1 Calculate the DFT of $x[n] = [1, 0, -1, 0]$

Solution 4.1

Let define N and label each entry of $x[n]$, before applying the analysis equation:
$N = 4$ (the number of samples in $x[n]$)

$$x[n] = \left[\overbrace{1}^{x[0]}, \overbrace{0}^{x[1]}, \overbrace{-1}^{x[2]}, \overbrace{1}^{x[3]}\right]$$

then, DFT of $x[n]$ can be derived as;

$$X[k] = \sum_{n=0}^{N-1} x[n]W_N^{kn} \quad ; k = 0, 1, \ldots, N-1$$

$$X[0] = x[0]W_N^0 + x[1]W_N^0 + x[2]W_N^0 + x[3]W_N^0$$

$$X[1] = x[0]W_N^0 + x[1]W_N^1 + x[2]W_N^2 + x[3]W_N^3$$

$$X[2] = x[0]W_N^0 + x[1]W_N^2 + x[2]W_N^4 + x[3]W_N^6$$

$$X[3] = x[0]W_N^0 + x[1]W_N^3 + x[2]W_N^6 + x[3]W_N^9$$

$$W_4^0 = e^{-j\frac{2\pi}{4}(0)} = 1\angle 0$$

$$W_4^1 = e^{-j\frac{2\pi}{4}(1)} = e^{-j\frac{\pi}{2}} = 1\angle -\pi/2$$

$$W_4^2 = e^{-j\frac{2\pi}{4}(2)} = e^{-j\pi} = 1\angle -\pi$$

$$W_4^3 = e^{-j\frac{2\pi}{4}(3)} = e^{-j\pi} = 1\angle -3\pi/2 = 1\angle \pi/2$$

$$W_4^4 = e^{-j\frac{2\pi}{4}(4)} = e^{-j\pi} = 1\angle -2\pi = 1\angle 0$$

$$W_4^6 = e^{-j\frac{2\pi}{4}(6)} = e^{-j\pi} = 1\angle -3\pi = 1\angle -\pi$$

$$W_4^9 = e^{-j\frac{2\pi}{4}(9)} = e^{-j\pi} = 1\angle -3\pi = 1\angle -9\pi/2 = 1\angle -\pi/2$$

Thereafter, DFT coefficients $X[k]$ are calculated as;

$$X[0] = 1$$
$$X[1] = 2 + j$$
$$X[2] = -1$$
$$X[3] = 2 + j$$

Magnitude an phase angle of $X[k]$ are shown in Fig. 4.1.

Taking into consideration the explanations discussed above, DFT calculations can be carried out by means of matrix algebra. For instance, the matrix form of DFT calculations of a discrete signal of length-4 can be written as;

$$\begin{bmatrix} X[0] \\ X[1] \\ X[2] \\ X[3] \end{bmatrix} = \begin{bmatrix} 1 & 1 & 1 & 1 \\ 1 & W_N & W_N^2 & W_N^3 \\ 1 & W_N^2 & W_N^4 & W_N^6 \\ 1 & W_N^3 & W_N^6 & W_N^9 \end{bmatrix} \begin{bmatrix} x[0] \\ x[1] \\ x[2] \\ x[3] \end{bmatrix}$$

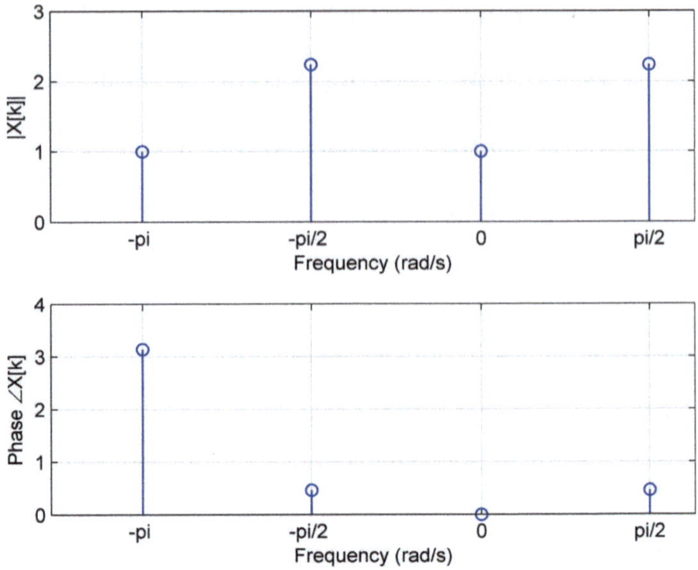

Fig. 4.1 Magnitude an phase angle of $X[k]$ in Example 4.1

In general notation the matrix equation can be denoted as;

$$\begin{bmatrix} X[0] \\ X[1] \\ X[2] \\ \vdots \\ X[N-1] \end{bmatrix} = \begin{bmatrix} W_N^0 & W_N^0 & W_N^0 & \cdots & W_N^0 \\ W_N^0 & W_N^1 & W_N^2 & \cdots & W_N^{N-1} \\ W_N^0 & W_N^2 & W_N^4 & \cdots & W_N^{2(N-1)} \\ \cdots & \cdots & \cdots & \cdots & \cdots \\ W_N^0 & W_N^{N-1} & W_N^{2(N-1)} & \cdots & W_N^{(N-1)^2} \end{bmatrix} \begin{bmatrix} x[0] \\ x[1] \\ x[2] \\ \vdots \\ x[N-1] \end{bmatrix}$$

or, in a compact form;

$$\mathbf{X} = \mathbf{W_N x}$$

Similar manner, using the inverse-DFT scheme,

$$\mathbf{x} = \mathbf{W_N^{-1} X}$$

where $\mathbf{W_N^{-1}}$ is the inverse of $\mathbf{W_N}$, and it is defined as;

$$\mathbf{W_N^{-1}} = \frac{1}{N}\mathbf{W_N^*}$$

$\mathbf{W_N^*}$ represents the complex conjugate of $\mathbf{W_N}$.

Example 4.2 Considering the discrete time signal $x[n] = \{1, 1, 1, 1\}$; investigate DTFT and DFT relation for various values of N.

Solution 4.2

Using Table 3.1, DTFT of $x[n]$ can be calculated as;

$$x[n] = \begin{cases} 1, & 0 \leq n \leq M \\ 0, & \text{otherwise} \end{cases} \leftrightarrow \frac{\sin\left[\omega\left(\frac{M+1}{2}\right)\right]}{\sin\left(\frac{\omega}{2}\right)} e^{-\frac{j\omega M}{2}}$$

where M is 4 in this example. The corresponding $|X\left(e^{j\omega}\right)|$, and $X[k]$ are depicted in Fig. 4.2. For $N \neq 4$, zero-padding approach is employed, i.e. $x[n] = \{1, 1, 1, 1, 0, 0, 0, 0\}$ for $N = 8$. One can easily see that $|X\left(e^{j\omega}\right)|$ is symmetric with respect to y-axis; and $X[k]$ has values only for non-negative ks. In addition, $X[k]$ can be considered as an appropriate samples of $|X\left(e^{j\omega}\right)|$.

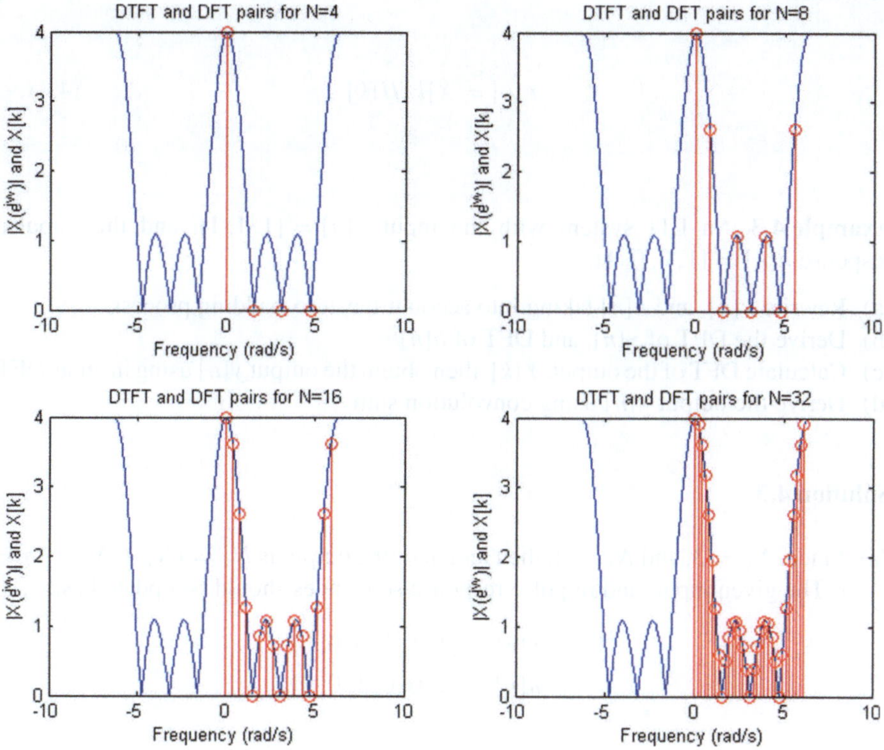

Fig. 4.2 DTFT and DFT pairs for various values of N, Example 4.2

4.2 LTI Systems Analysis Using DFT

Considering a discrete LTI system, the relationship between the input and the output, $y[n]$, can be expressed by employing convolution sum of the input, $x[n]$, and the impulse response of the system, $h[n]$, such as;

$$y[n] = x[n] * h[n]$$

Assuming that, both input and the impulse response are finite-length sequences with the lengths of N_x, and N_h, respectively. Therefore, the convolution sum results in the output, $y[n]$, having the length of $(N_x + N_h - 1)$. In order to employ DFT for LTI system analysis, the lengths of $x[n]$, and $h[n]$ must be the same length with the output, $y[n]$. Hence, their lengths will be completed to $(N_x + N_h - 1)$ by adding needed number of zeros, this operation is named as zero-padding. Thereafter, it can be written as;

$$Y[k] = X[k]H[k] \tag{4.3}$$

Example 4.3 An LTI system with the input $x[n] = [1, 1, 1]$, and the impulse response $h[n] = [1, 0, 0, 1]$,

(a) Rewrite $x[n]$, and $h[n]$ taking into account the zero-padding process
(b) Derive the DFT of $x[n]$, and DFT of $h[n]$
(c) Calculate DFT of the output, $Y[k]$, then obtain the output $y[n]$ using inverse-DFT
(d) Derive the output $y[n]$ using convolution sum

Solution 4.3

(a) Since, $N_x = 3$, and $N_h = 4$, then length of the output is $N_y = (N_x + N_h - 1) = 6$ The given input, and impulse response sequences should be updated as;

$$x[n] = [1, 1, 1, 0, 0, 0]$$
$$h[n] = [1, 0, 0, 1, 0, 0]$$

(b) $X[k]$, and $H[k]$ is obtained using Analysis equation

$$X[k] = \sum_{n=0}^{N-1} x[n]e^{-j\frac{2\pi}{N}kn} \quad \text{Analysis Equation}$$

where, $N = 6$. The results are listed as

$$X[k] = \left[3, 1 - j\sqrt{3}, 0, 1, 0, 1 + j\sqrt{3}\right]$$
$$H[K] = [2, 0, 2, 0, 2, 0]$$

(c) DFT of the output is calculated as;

$$Y[k] = X[k]H[k] = [6, 0, 0, 0, 0, 0]$$

The output in time domain, $y[n]$, is calculated using Synthesis equation as;

$$y[n] = \frac{1}{N}\sum_{k=0}^{N-1} Y[k]e^{-j\frac{2\pi}{N}kn}$$
$$= [1, 1, 1, 1, 1, 1]$$

(d) The output can be derived by means of the convolution sum in time domain as;

$$y[n] = \sum_{k=-\infty}^{\infty} x[k]h[n-k]$$
$$= [1, 1, 1, 1, 1, 1]$$

which is exactly the same the result obtained in the previous part.

4.3 Fast Fourier Transform (FFT)

DFT can be considered as a proper tool in order to be employed in digital signal processing applications due to its discrete nature. $x[n]$ is a discrete-time signal, and its corresponding DFT pair $X[k]$ is also discrete in frequency domain. However, there still is a significant problem remains waiting to be solved in application layer: the computational cost. As discussed in the previous section, there should be N^2 complex multiplication, and $N(N-1)$ complex addition have to be carried out in order to complete DFT calculations. Number of computations in DFT calculations are shown in Table 4.1 as illustrative purpose.

Table 4.1 Number of computations in a DFT calculation

N	Complex multiplication	Complex addition	Total
2	4	2	6
4	16	12	28
8	64	56	120
16	256	240	496
32	1024	992	2016
64	4096	4032	8128
128	16384	16256	32640
256	65536	65280	130816
512	262144	261632	523776
1024	1048576	1047552	2096128
2048	4194304	4192256	8386560
4096	16777216	16773120	33550336
8192	67108864	67100672	1.34E+08

The discrete Fourier transform (DFT) of a sequence is efficiently computed using the FFT algorithm. The DFT is a mathematical transformation that converts a time-domain signal into a frequency-domain signal. In digital signal processing and communications systems, the FFT is widely used. Please note that, FFT is not a new transformation technique, it is just a fast algorithm in order to compute the DFT in a fast and more efficient way. There are many approaches to build up an FFT algorithm in literature, we will adopt the following one, described in Algorithm 1, for the sake of better understanding and simplicity.

Hereafter the simplification procedures, the order of operations in FFT can be given as $Nlog_2(N)$, while the order of operations is N^2 in DFT. Because complex arithmetic, especially multiplication, is very costly, it thus represents a significant cost saving over DFT notably for $N > 100$. A computational cost comparison between DFT and FFT is given in Table 4.2.

FFT is widely used in digital signal processing for a variety of applications, including: *Spectrum analysis*: It provides a frequency-domain representation of a signal and is used to identify its frequency components. *Filter design*: It can be employed to create low-pass, high-pass, and band-pass filters. *Image processing*: It can be utilized to compress and denoise images, as well as extract features and recognize images. It plays a role in *audio processing*, such as equalization, audio compression, and noise reduction. *Communication systems:* It is used for modulation and demodulation, channel equalization, and signal detection in digital communication systems. *Medical imaging*: It is used to extract useful information from medical imaging, such as MRI and CT scans.

A signal flow graph might be accompany calculations as an illustrative tool in order to improve the understanding of the topic. An 8-point FFT signal flow graph is

Algorithm 1 Developing an FFT algorithm

1. The length of the sequence, N must be power of 2, i.e. $N = 2^K$ where $K = 1, 2, 3, \ldots$
2. If $N \neq 2^K$, then complete the sequence by means of zero-padding to the next integer which is power of 2
3. Split the Analysis Equation into two parts, such as even-indexed, and odd-indexed values

$$\begin{aligned} X[k] &= \sum_{n=0}^{N/2-1} x[2n]W_N^{2nk} + \sum_{n=0}^{N/2-1} x[2n+1]W_N^{(2n+1)k} \\ &= \sum_{n=0}^{N/2-1} x(2n)W_N^{2nk} + W_N^k \sum_{n=0}^{N/2-1} x(2n+1)W_N^{2nk} \end{aligned}$$

Please note that, W_N^{2nk} can be expressed as follows;

$$\begin{aligned} W_N^{2nk} &= e^{-j2\pi(2nk)/N} \\ &= e^{-j2\pi(nk)/(N/2)} \\ &= W_{N/2}^{nk} \end{aligned}$$

Thereafter, the DFT coefficients $X[k]$ can be calculated as;

$$X[k] = \sum_{n=0}^{N/2-1} x[2n]W_{N/2}^{nk} + W_N^k \sum_{n=0}^{N/2-1} x[2n+1]W_{N/2}^{nk}$$

The previous explanation indicates that N-length DFT might be calculated using 2 of $\frac{N}{2}$-length DFTs.

4. Proceed the same procedure until getting $N = 2$, i.e. $x[n] = [x[0], x[1]]$
5. Use the following simplifications

- If $n = 0$

$$W_N^{nk} = e^{-j2\pi(2nk)/N} = 1$$

- If $n = \frac{N}{4}$

$$\begin{aligned} W_N^{nk} &= e^{-j2\pi(2(N/4)k)/N} \\ &= e^{-j\pi/2k)} \\ &= (-j)^k \end{aligned}$$

- If $n = \frac{N}{2}$

$$\begin{aligned} W_N^{nk} &= e^{-j2\pi(2(N/2)k)/N} \\ &= e^{-j\pi k)} \\ &= (-1)^k \end{aligned}$$

- If $n = \frac{3N}{4}$

$$\begin{aligned} W_N^{nk} &= e^{-j2\pi(2(3N/4)k)/N} \\ &= e^{-j3\pi/2k)} \\ &= (j)^k \end{aligned}$$

Table 4.2 Computational cost comparison between DFT and FFT

	DFT	FFT	FFT/DFT
N	N^2	$Nlog_2(N)$	%
8	64	24	37.5000
16	256	64	25.0000
32	1024	160	15.6250
64	4096	384	9.3750
128	16384	896	5.4688
256	65536	2048	3.1250
512	261144	4608	1.7645
1024	1048576	10240	0.9766
2048	4194304	22528	0.5371
4096	16777216	49152	0.2930

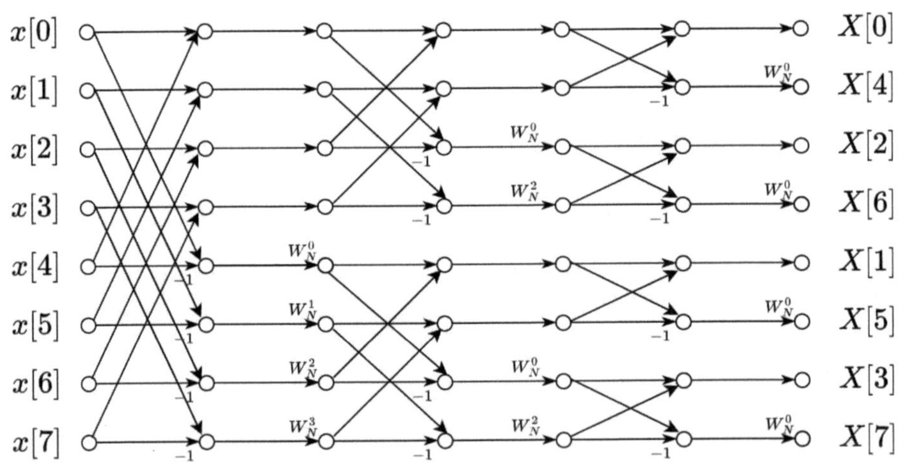

Fig. 4.3 Signal flow graph for 8-point FFT calculations

depicted in Fig. 4.3. Contributions of each components of the sequence might easily be seen by following the signal flows.

Example 4.4 In Fig. 4.4, FFT signal flow graph is given where $N = 8$. The bold line shows the flow from $x[1]$ to DFT sample $X[5]$.

(a) Calculate the gain following the bold line, i.e. $X[5] = \text{Gain}x[1]$
(b) Find $X[5]$ using;

$$X[5] = \sum_{n=0}^{N-1} x[n]e^{-j\frac{2\pi}{N}n(5)}$$

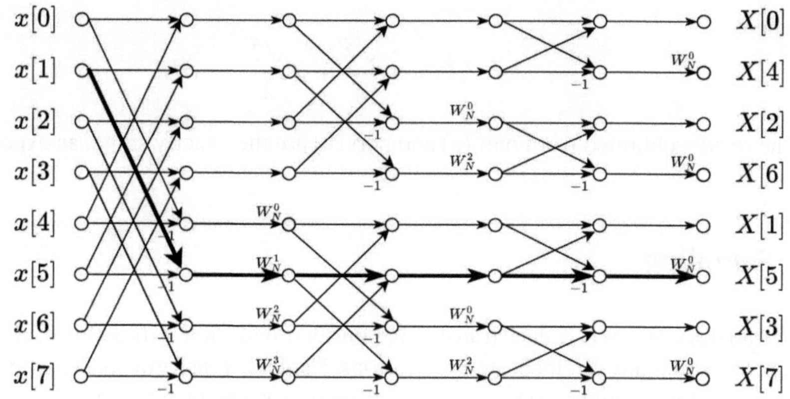

Fig. 4.4 Signal flow graph for 8-point FFT calculations, Example 4.4

(c) Compare the results obtained from the previous parts, is the contribution of $x[1]$ in $X[5]$ same, or?

Solution 4.4

(a) From the graph, it can be calculated as

$$X[5] = \underbrace{\left(w_N^1\right)(-1)\left(w_N^0\right)}_{\text{Gain}} x[1]$$

$$W_N^1 = e^{-j\frac{2\pi}{8}(1)} = e^{-j\frac{\pi}{4}} = \frac{\sqrt{2}}{2} - j\frac{\sqrt{2}}{2}$$

$$W_N^0 = e^{-j\frac{2\pi}{8}(0)} = 1$$

Therefore, the Gain is

$$Gain = \left(\frac{\sqrt{2}}{2} - J\frac{\sqrt{2}}{2}\right)(-1)(1)$$

$$= -\frac{\sqrt{2}}{2} + J\frac{\sqrt{2}}{2}$$

(b)

$$x[5] = x[1]e^{-5\frac{2\pi}{8}(1)(5)} = e^{-j\frac{5\pi}{4}}x[1]$$

where

$$e^{-j\frac{5\pi}{4}} = -\frac{\sqrt{2}}{2} + J\frac{\sqrt{2}}{2}$$

(c) The results obtained from part (a) and part (b) are the exactly same, as expected.

4.4 Sampling

Linear transformations (Laplace transform, Fourier transforms, Z-transform, etc.) are efficient tools in many engineering applications. Laplace transform can be employed in continuous time signal processing applications under certain region of convergence conditions. For instance it is used in order to analyse stability and causality properties of such systems. A general overview about linear transformations can be seen in Table 4.3.

In digital signal processing, DFT is employed in order to design, understand, and analysis of systems by means of FFT. Therefore, sampling should be taking into consideration as a preprocessing layer aiming to express the related signals in discrete-time nature.

Sampling is a process in signal processing where a continuous signal is converted into a sequence of discrete values, representing the amplitude of the signal at regular intervals. The output of the sampling process is a discrete-time signal.

The basic idea behind sampling is to represent a continuous signal by a sequence of discrete values that are spaced at equal intervals in time. The rate at which the signal is sampled is known as the sampling rate or sampling frequency, and it is usually expressed in samples per second (Hz).

Table 4.3 Common transformation tools in signal processing area

Signal	Transformation tool
Continuous-time	Laplace Transform
Continuous-time, periodic	Fourier Series
Continuous-time, aperiodic	Fourier Transform
Discrete-time	Z-Transform
Discrete-time periodic	Discrete Fourier Series (DFS)
Discrete-time aperiodic	Discrete Time Fourier Transform (DTFT)
Discrete-time finite sequence	Discrete Fourier Transform (DFT)/FFT algorithm

The sampling theorem states that a continuous signal can be perfectly reconstructed from its sampled values if the sampling rate is greater than twice the highest frequency component in the signal. This is known as the Nyquist rate. If the sampling rate is lower than the Nyquist rate, the sampled signal will contain aliasing artifacts, which are distortions in the reconstructed signal caused by the overlap of frequency components.

Sampling is a fundamental operation in digital signal processing, as it provides a convenient way to process signals using digital computers. The sampled signal can be stored, transmitted, and processed digitally, allowing for efficient and accurate analysis and manipulation of signals.

A periodic pulse train signal is a type of signal that consists of a sequence of pulses, or rectangular waveform segments, repeated at regular intervals. The time between each repetition of the pulse sequence is known as the period of the signal.

A periodic pulse train signal can be described by its pulse width, which is the duration of the pulse, and its duty cycle, which is the fraction of each period that is occupied by the pulse. The duty cycle can be used to control the average value of the signal, which can be useful in many applications, such as control systems and communication systems.

Periodic pulse train signals are commonly used as clock signals in digital circuits, as they provide a simple and reliable way to synchronize events. They are also used as input signals in testing and simulation, as well as in some control systems, where they are used to generate a series of pulses with a specific frequency and pulse width.

Let define the periodic pulse train signal as;

$$p(t) = \sum_{n=-\infty}^{\infty} \delta(t - nT_s) \tag{4.4}$$

where $\delta(t)$ is the unit impulse function, and T_s is the sampling period.
Assuming that the representation of a continuous-time signal,$x(t)$, can be obtained as;

$$\begin{aligned} x_s(t) &= x(t)p(t) \\ &= x(t) \sum_{n=-\infty}^{\infty} \delta(t - nT_s) \\ &= \sum_{n=-\infty}^{\infty} x(nT_s)\, \delta(t - nT_s) \end{aligned} \tag{4.5}$$

Fourier transform of the periodic impulse train, $p(t)$ is

$$P(j\omega) = \frac{2\pi}{T_s} \sum_{k=-\infty}^{\infty} \delta(\omega - k\omega_s)$$

where ω_s is the sampling frequency, and

$$\omega_s = \frac{2\pi}{T_s}$$

Thereafter, the Fourier transform of the sampled signal, $x_s(t)$ can be obtained employing convolution in frequency domain as;

$$\begin{aligned} X_s(j\omega) &= \frac{1}{2\pi}\left[(X(j\omega)) * \left(\frac{2\pi}{T_s}\sum_{k=-\infty}^{\infty}\delta\left(j(\omega - k\omega_s)\right)\right)\right] \\ X_s(j\omega) &= \frac{1}{T_s}\sum_{k=-\infty}^{\infty} X\left(j(\omega - k\omega_s)\right) \end{aligned} \tag{4.6}$$

A band-limited signal in the frequency domain is a signal that has its spectral energy concentrated within a specific frequency range, or band. In other words, the signal has limited bandwidth, which means that its frequency components are confined to a certain interval of frequencies.

The bandwidth of a band-limited signal is typically defined as the difference between the highest and lowest frequencies present in the signal. Band-limited signals are important in many applications, including communication systems, where the signal must be transmitted over a limited frequency band, and image and audio processing, where it is desirable to represent the signal with a limited number of frequency components.

In the frequency domain, a band-limited signal can be represented as a sum of sinusoidal components, with each component having a specific amplitude and phase. By restricting the number of frequency components, the signal can be efficiently processed and represented using digital techniques.

The concept of band-limiting a signal is closely related to the Nyquist-Shannon sampling theorem, which states that a continuous signal can be perfectly reconstructed from its samples if the signal is band-limited and the sampling rate is greater than twice the signal's bandwidth.

A band-limited signal representation in frequency domain is shown in Fig. 4.5. Please note that $X(j\omega) = 0$ apart from $-\omega_m < \omega < \omega$.

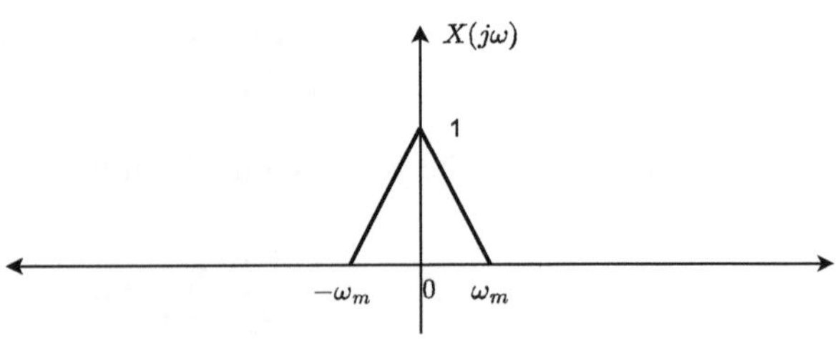

Fig. 4.5 Frequency domain representation of a band limited signal

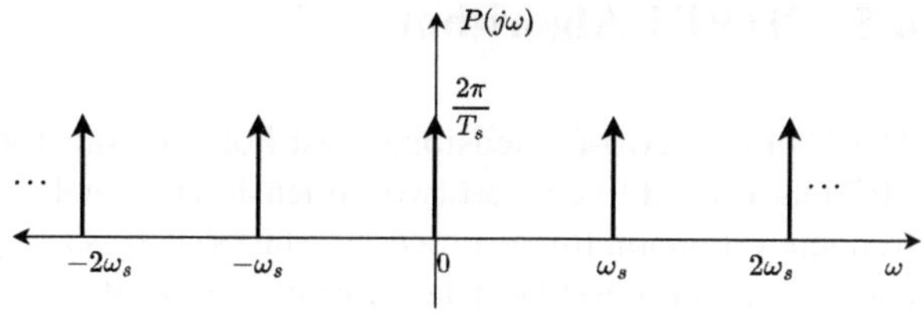

Fig. 4.6 The Fourier transform of the sampling signal

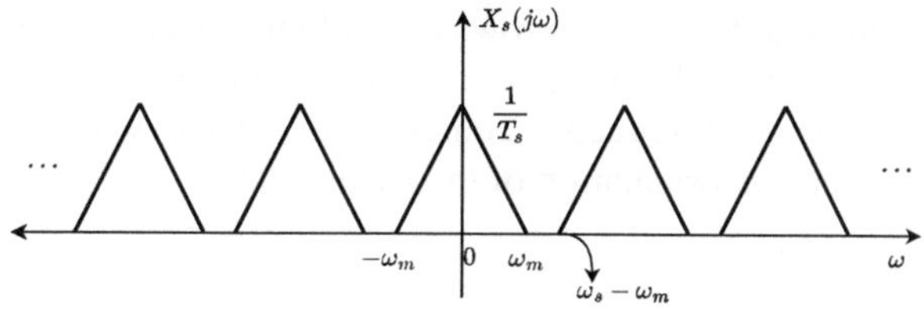

Fig. 4.7 The Fourier transform of the sampled signal, $\omega_s > 2\omega_m$

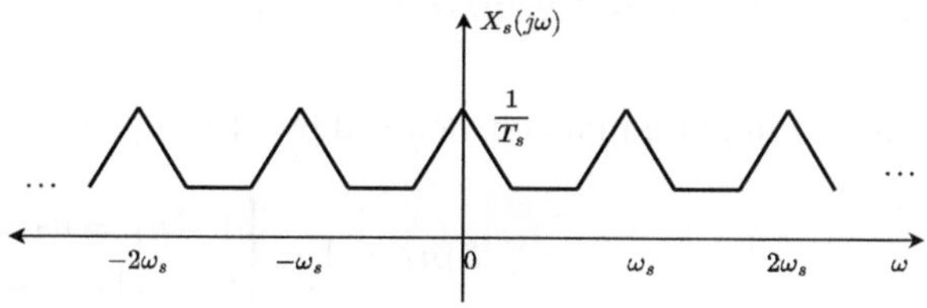

Fig. 4.8 The Fourier transform of the sampled signal, $\omega_s < 2\omega_m$

The Fourier transform of the sampling signal, which is an impulse train, is given in Fig. 4.6.

The Fourier transform of the sampled signal where $\omega_s > 2\omega_m$ is illustrated in Fig. 4.7.

Aliasing is an effect in signal processing that occurs when a high-frequency signal is sampled at a rate that is lower than the Nyquist rate. The Nyquist rate is defined as twice the highest frequency present in the signal.

When a signal is sampled at a rate lower than the Nyquist rate, some of the higher frequency components in the signal are indistinguishable from lower frequency components, leading to a phenomenon known as aliasing. This occurs because the samples do not contain enough information to accurately represent the higher frequency components in the signal.

In the frequency domain, aliasing can be seen as the overlapping of frequency components from different parts of the signal spectrum, resulting in a distorted representation of the signal. The distorted signal will contain both the original signal and additional components that were not present in the original signal, known as alias components.

Aliasing can be avoided by applying anti-aliasing filters to the signal before sampling, which remove the high-frequency components above the Nyquist rate. This ensures that the signal can be accurately represented using a lower sampling rate. The Fourier transform of the sampled signal where $\omega_s < 2\omega_m$ is illustrated in Fig. 4.8.

4.5 2D FFT Algorithm

The 2D FFT (Two-Dimensional Fast Fourier Transform) is an extension of the 1D FFT that is used to convert two-dimensional signals from the spatial domain to the frequency domain. Image processing tasks such as image filtering, feature extraction, and image compression make extensive use of it.

The input signal to a 2D FFT is a 2D matrix representing an image. The FFT is applied to the image's rows and columns, resulting in a matrix of complex numbers representing the image's frequency components. The magnitude of these numbers indicates the frequency content of the image, while the phase information indicates the relative orientation of the frequencies.

4.5.1 2D Unit Impulse Function

The $2D$ impulse function, denoted by $\delta[n_1, n_2]$, is defined as

$$\delta[n_1, n_2] = \begin{cases} 1, & n_1 = n_2 = 0 \\ 0, & \text{otherwise} \end{cases} \tag{4.7}$$

A $2D$ sequence $x[n_1, n_2]$ can be expressed as a linear combination of impulse functions as follows:

$$x[n_1, n_2] = \sum_{k_1=-\infty}^{\infty} \sum_{k_2=-\infty}^{\infty} x[k_1, k_2]\, \delta[n_1 - k_1, n_2 - k_2] \tag{4.8}$$

$\delta[n_1, n_2]$ is a separable function due to $\delta[n_1, n_2]$ can be written as

$$\delta[n_1, n_2] = \delta[n_1]\, \delta[n_2]$$

where $\delta[n_1]$ and $\delta[n_2]$ are $1D$ unit impulse functions.

4.5.2 2D Unit Step Function

The $2D$ unit step function, denoted by $u[n_1, n_2]$, is expressed as

$$u[n_1, n_2] = \begin{cases} 1, & n_1, n_2 \geq 0 \\ 0, & \text{otherwise} \end{cases} \tag{4.9}$$

The sequence $u[n_1, n_2]$ can be explained related to $\delta[n_1, n_2]$ as

$$u[n_1, n_2] = \sum_{k_1=-\infty}^{n_1} \sum_{k_2=-\infty}^{n_2} \delta[k_1, k_2] \tag{4.10}$$

On the other hand, one can easily derive the unit impulse function in terms of the unit step function as follows;

$$\delta[n_1, n_2] = u[n_1, n_2] - u[n_1 - 1, n_2] - u[n_1, n_2 - 1] + u[n_1 - 1, n_2 - 1] \tag{4.11}$$

4.5.3 Systems

Considering a one-to-one mapping from any given input to a certain output, then the input-output mapping is called a system, which is denoted as $\mathrm{T}\{\cdot\}$. $\mathrm{T}\{\cdot\}$ maps an input $x[n_1, n_2]$ to an output $y[n_1, n_2]$ is shown as;

$$y[n_1, n_2] = \mathrm{T}\{x[n_1, n_2]\}$$

If the input is 2D unit impulse function, i.e. $\delta[n_1, n_2]$, then the corresponding output is called the impulse response; by definition

$$h[n_1, n_2] = \mathrm{T}\{\delta[n_1, n_2]\}$$

4.5.4 Convolution

For LTI systems, the relationship between the input and the corresponding output can be defined by means of convolution sum. Let the output be;

$$\begin{aligned} y[n_1, n_2] &= \mathrm{T}\{x[n_1, n_2]\} \\ &= \mathrm{T}\left\{\sum_{k_1=-\infty}^{\infty} \sum_{k_2=-\infty}^{\infty} x[k_1, k_2]\,\delta[n_1 - k_1, n_2 - k_2]\right\} \\ &= \sum_{k_1=-\infty}^{\infty} \sum_{k_2=-\infty}^{\infty} x[k_1, k_2]\,\mathrm{T}\{\delta[n_1 - k_1, n_2 - k_2]\} \\ &= \sum_{k_1=-\infty}^{\infty} \sum_{k_2=-\infty}^{\infty} x[k_1, k_2]\,h[n_1 - k_1, n_2 - k_2] \end{aligned} \tag{4.12}$$

which is called convolution sum, and simply it can be expressed as;

$$y[n_1, n_2] = x[n_1, n_2] * h[n_1, n_2] \tag{4.13}$$

4.5.5 2D DFT Pairs

A straightforward extension of the 1-D result yields the 2D DFT representation. In particular, the 2D DFT pair is given by

$$X[k_1, k_2] = \begin{cases} \sum_{n_1=0}^{N_1-1} \sum_{n_2=0}^{N_2-1} x[n_1, n_2] e^{-j(2\pi/N_1)k_1 \cdot n_1} e^{-j(2\pi/N_2)k_2 \cdot n_2}, & \text{for } 0 \le k_1 \le N_1 - 1, \\ & 0 \le k_2 \le N_2 - 1 \\ 0, & \text{otherwise} \end{cases}$$

$$x[n_1, n_2] = \begin{cases} \frac{1}{N_1 N_2} \sum_{k_1=0}^{N_1-1} \sum_{k_2=0}^{N_2-1} X[k_1, k_2] e^{j(2\pi/N_1)k_1 n_1} e^{j(2\pi/N_2)k_2 n_2}, & \text{for } 0 \le n_1 \le N_1 - 1, \\ & 0 \le n_2 \le N_2 - 1 \\ 0 & \text{otherwise} \end{cases} \tag{4.14}$$

4.5.6 2D FFT Algorithm

The discrete Fourier transform (DFT) of a two-dimensional signal is computed using the 2D Fast Fourier Transform (FFT). It brilliantly engages the 1D FFT algorithm to reduce the computational complexity of DFT. The two-dimensional fast Fourier transform (FFT) is widely used in image processing, signal processing, and other fields for tasks such as image filtering, convolution, and correlation.

The 2D FFT can be broken down into two 1D FFTs, applied to the rows and columns of the input matrix. Therefore, computational cost of the 2D FFT can be dramatically reduced. The algorithm starts with rearranging the data in the input matrix, the rows and columns are transformed separately. The 1D FFT is applied to each row, and then the result is transposed. After the transposition, the 1D FFT is applied to each column of the transposed matrix. The final result of the 2D FFT is obtained by transposing the result of the second 1D FFT back to its original form.

In image processing, the 2D FFT is used to perform various operations in the frequency domain. Some common applications include:

Image filtering: The 2D FFT is used to apply filters to images, such as high-pass, low-pass, and band-pass filters. This allows you to remove or enhance certain frequency components in an image.
Edge detection: The 2D FFT can be used for edge detection by finding the high-frequency components in the frequency domain representation of an image.
Image compression: The 2D FFT can be used for image compression by removing low-frequency components from the frequency domain representation of an image. This results in a smaller representation of the image with less information loss.
Image restoration: The 2D FFT can be used for image restoration by removing noise from an image in the frequency domain. This is done by setting certain frequency components to zero or modifying them in some way.
Image deblurring: The 2D FFT can be used for image deblurring by restoring the original image from a blurred version in the frequency domain. This is done by modifying the frequency components in the blurred image.

The ability to perform operations in the frequency domain makes the 2D FFT a powerful tool for image processing tasks.

4.6 Some Applications of FFT

4.6.1 Sinusoidal Sequence

Consider a discrete time sinusoidal sequence given as;

$$x[n] = \cos(2\pi 5n - \pi/2) - \sin(2\pi 5n + \pi/2)$$

The corresponding DFT coefficients are shown in Fig. 4.9.

Problems

4.1 Explain why the Discrete Time Fourier Transform (DTFT) might be inadequate for direct implementation in digital signal processing systems, despite its analytical significance.

4.2 Define the Discrete Fourier Transform (DFT) of a finite-length discrete signal $x[n]$ of length N. Write down the analysis and synthesis equations for the DFT.

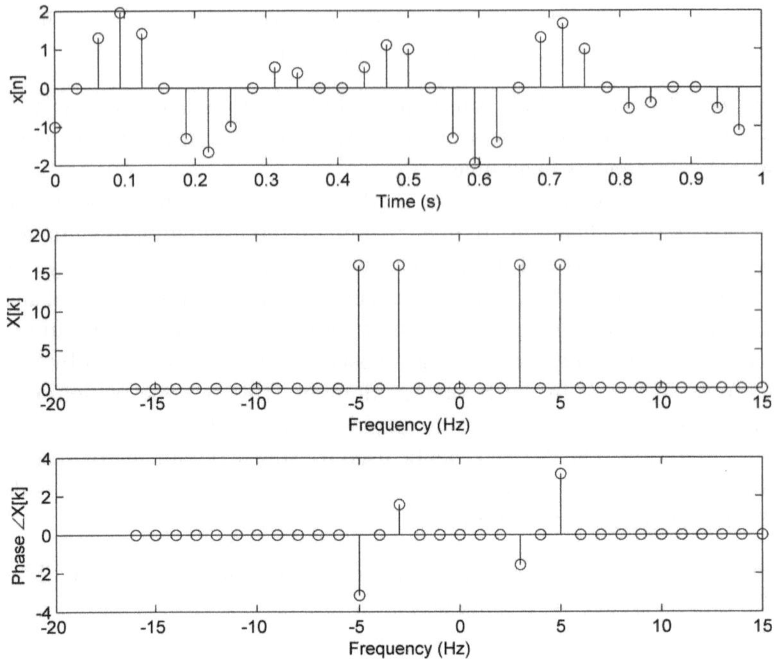

Fig. 4.9 Magnitude an phase angle of $X[k]$ of a sinusoidal signal

4.3 What is the twiddle factor W_N? Express it mathematically and explain its role in the DFT computation.

4.4 Given the discrete signal $x[n] = [2, 1, 0, -1]$, calculate its 4-point DFT, $X[k]$, using the analysis equation.

4.5 Describe how the DFT of a finite-length sequence can be computed using matrix algebra. Write down the general matrix form for an N-point DFT.

4.6 What is the relationship between the DFT matrix $\mathbf{W_N}$ and its inverse $\mathbf{W_N^{-1}}$? How is the inverse DFT computed using the DFT matrix?

4.7 In Example 4.2, how does zero-padding the time-domain signal $x[n]$ affect its DFT $X[k]$ in terms of the number of frequency samples and their relationship to the DTFT?

4.8 Explain how the DFT can be used to analyze Linear Time-Invariant (LTI) systems. What operation in the frequency domain corresponds to convolution in the time domain? What condition must be met regarding the lengths of the input and impulse response sequences?

4.9 In Example 4.3, why was it necessary to zero-pad the input $x[n]$ and the impulse response $h[n]$ before computing their DFTs for LTI system analysis? What is the length of the resulting output $y[n]$ in both the time and frequency domains?

4.10 Compare the computational complexity of directly computing an N-point DFT with that of an N-point Fast Fourier Transform (FFT). What is the order of operations for each? Why is the FFT considered a "fast" algorithm?

4.11 State the key requirement for the length N of a sequence when using the basic FFT algorithm described in Algorithm 1. What is done if the sequence length does not meet this requirement?

4.12 Briefly outline the steps involved in the divide-and-conquer approach used in the FFT algorithm as described in Algorithm 1. How does it reduce the number of computations?

4.13 In the 8-point FFT signal flow graph (Fig. 4.3), what do the nodes and directed edges represent? What is the significance of the twiddle factors associated with the edges?

4.14 In Example 4.4, explain how the gain from a specific input sample $x[1]$ to a specific DFT output sample $X[5]$ is determined by following the signal flow graph. How does this relate to the direct DFT calculation?

4.15 Define the process of sampling a continuous-time signal. What is the sampling rate or sampling frequency, and what are its units?

4.16 State the Nyquist-Shannon sampling theorem. What is the Nyquist rate, and why is it important for perfect signal reconstruction?

4.17 Explain the phenomenon of aliasing in the context of signal sampling. What causes it, and how does it manifest in the frequency domain? How can aliasing be avoided?

4.18 Describe the 2D Discrete Fourier Transform (DFT) for a 2D signal $x[n_1, n_2]$ of size $N_1 \times N_2$. Write down the 2D DFT pair equations.

4.19 How is the 2D Fast Fourier Transform (FFT) algorithm typically implemented? How does it leverage the 1D FFT algorithm to efficiently compute the 2D DFT?

4.20 List and briefly describe at least three applications of the FFT in digital signal processing, image processing, or communication systems mentioned in the chapter.

Chapter 5
Random Variables

In engineering and science, the experimental or measured/collected data has a significant role in many applications. Even though, the nature of the investigated data may have continuous form while mentioning about the measurement, the obtained data must be in discrete form due to, at least, sensitivity level of the measurement tools. For instance, the temperature in a room might be between 10 and 40 °C. In theory, the temperature may have any value within this interval. However, when one measures the temperature at various times, the results will be in discrete form. As a result of this, the random variable X frequently needs to be assigned in a discrete form in many applications due to the nature of the problem being investigated.

5.1 The Probability Distribution Function (PDF)

The probability distribution functions of discrete random variables are stair-case in nature, consisting of combinations that jump at specific values and stay constant at other values. Figure 5.1 depicts a graph for illustration purposes.

The probability operation and $F_X(x)$ have the following relationship:

$$P(X \leq a) = F_X(a) \quad \text{for any } a \text{ such that } -\infty < a < \infty \tag{5.1}$$

Properties of $F_X(x)$

1. $F_X(x)$ is bounded between 0 and 1, then $0 \leq F_X(x) \leq 1$ for all x values.
2. Due to $F_X(x)$ is a cumulative function, therefore it is a nondecreasing function,

$$\text{If } x_1 < x_2, \text{thereafter, } F_X(x_1) \leq F_X(x_2)$$

M. Catak and T. Allahviranloo, *Discrete Random Processes*, Studies in Systems, Decision and Control 644, https://doi.org/10.1007/978-3-032-08630-3_5

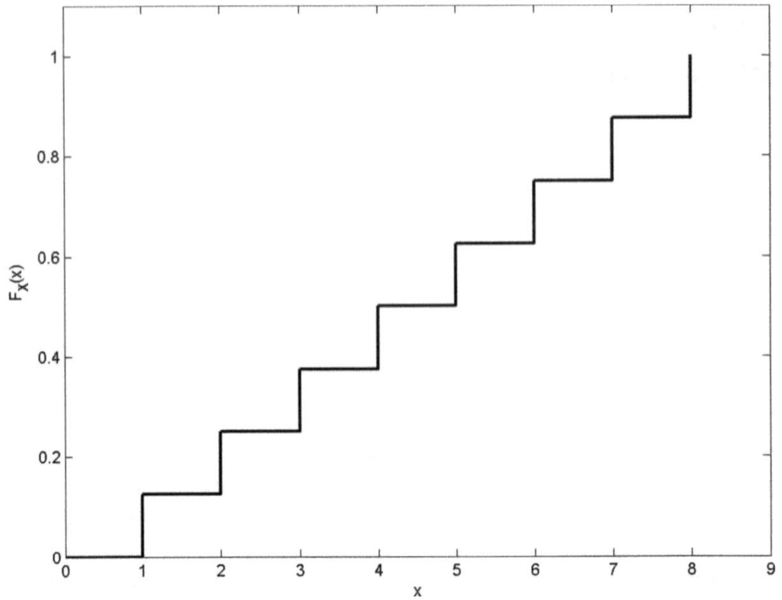

Fig. 5.1 A sample discrete probability distribution function for illustration purposes

3. The probability of the universal set, which includes all possibilities, is 1. $\lim_{x\to\infty} F_X(x) = 1$
4. The probability of the null set is 0. $\lim_{x\to-\infty} F_X(x) = 0$

5.2 The Probability Density Function (PDF)

The probability density function (pdf) stays at zero apart from some specific values at where the value of the distribution function changes. It can be defined as;

$$f_X(x) = \lim_{\epsilon\to 0} F_X(x+\epsilon) - F_X(x-\epsilon) \tag{5.2}$$

Properties of $f_X(x)$

1. Since $F_X(x)$ is a non-decreasing function, $f_X(x)$ must have a non-negative value
$$f_X(x) \geq 0 \quad \text{for all values of } x$$
2. Since, $\lim_{x\to\infty} F_X(x) = 1$, sum of $f_X(x)$ over its domain must be 1.
$$\sum_{\forall x} f_X(x_i) = 1$$
3. The distribution function $F_X(x)$ can be calculated as
$$F_X(x) = \sum_{x_i \leq x} f_X(x_i)$$
4. The probability of x having a certain value of a,
$$P(X = a) = \lim_{\epsilon\to 0} F_X(a+\epsilon) - F_X(a-\epsilon) = f_X(a)$$

Example 5.1 Let the probability density function be defined as;

$$f_X(x) = \sum_{x_i=1}^{\infty} A\left(\frac{1}{3}\right)^{x_i} \delta[x - x_i]$$

(a) Find A such that $f_X(x)$ is a valid probability density function.
(b) Find the probabilities of $P(1 < X \leq 3)$, $P(X > 2)$, $P(X = 3)$

Solution 5.1

(a) The following requirement is fulfilled by an adequate probability density function:

$$\textstyle\sum_{\forall x} f_X(x_i) = 1$$

$$\sum f_X(x_i) = \sum_{x_i=1}^{\infty} A\left(\frac{1}{3}\right)^{x_i} = A\frac{\frac{1}{3}}{1-\frac{1}{3}} = 1$$

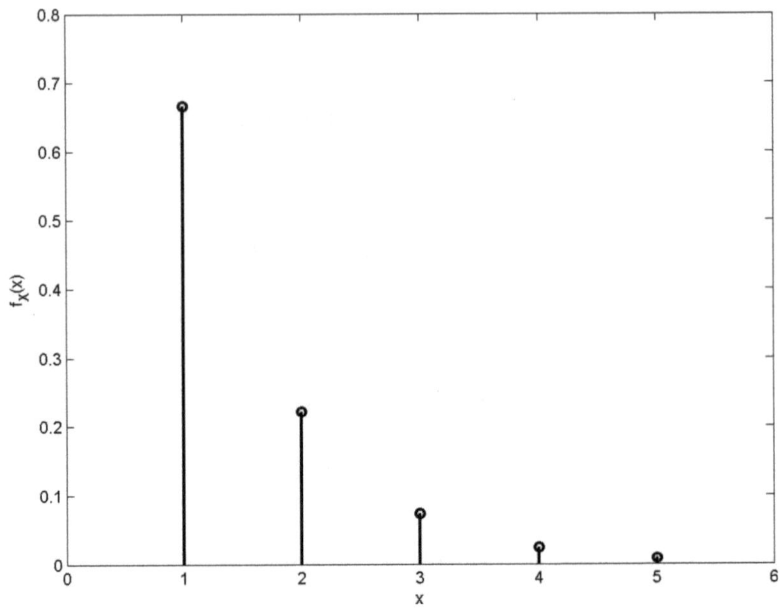

Fig. 5.2 Discrete probability density function (Example 5.1)

Hence, $A = 2$ in order to guarantee that $f_X(x)$ is a proper probability density function.

(b) Find the probabilities of $P(1 < X \leq 3)$, $P(X > 2)$, $P(X = 3)$

$$P(1 < X \leq 3) = 2\left(\frac{1}{3}\right)^2 + 2\left(\frac{1}{3}\right)^3 = \frac{4}{27} \approx 0.148$$

$$\begin{aligned} P(X > 2) &= 1 - P(X \leq 2) \\ &= 1 - 2\frac{1}{3} - 2\left(\frac{1}{3}\right)^2 = \frac{1}{9} \approx 0.111 \end{aligned}$$

$$P(X = 3) = 2\left(\frac{1}{3}\right)^3 = \frac{2}{27} \approx 0.704$$

The probability density function $f_X(x)$ is depicted in Fig. 5.2.

5.3 Expected Value and Variance

The variance, $Var[X]$, and the expected value, $E[X]$, are two important factors in comprehending a distribution's characteristics. The mean, μ_X, and the expected value are frequently used interchangeably. However, there is a minor distinction between the two: the expected value is only used for a distribution, but the mean is used to describe the average value of both a distribution and randomly generated data based on a probability distribution. Computing a probability distribution's expected value is done as follows:

$$
\begin{aligned}
E[X] &= \mu_X = \sum_{\forall x} x_i f_X(x_i) \\
\text{Var}[X] &= \sigma_X^2 = \sum_{\forall x} (x_i - \mu_X)^2 f_X(x_i)
\end{aligned} \tag{5.3}
$$

The corresponding random variable's likelihood of varying around the expected value is indicated by the non-negative parameter $Var[X]$. The variance squared is equal to the standard deviation, or σ_X. Rather than using variance for describing the probability density functions, σ_X is frequently employed.

$$\sigma_X = \sqrt{Var[X]}$$

5.4 Common Discrete Probability Distribution Functions

5.4.1 Uniform Distribution

If there is a finite number of x such that X has equally likely probability, then the discrete random variable X is said to be uniformly distributed. The expression for the related probability density function;

$$f_X(x) = \sum_{i=1}^{N} \frac{1}{N} \delta(x - x_i) \tag{5.4}$$

Assume that the definition of the uniform distribution lies in the interval $[k_1, k_2]$. Thus, $x_1 = k_1$, and $x_n = k_2$ are equivalent. Then, $N = k_2 - k_1 - +1$ is the number of locations in which X has a positive probability. For a uniform distribution, the expected value is computed as follows;

$$
\begin{aligned}
E[X] &= \sum_{x_i=k_1}^{k_2} x_i \frac{1}{k_2 - k_1 + 1} = \frac{1}{k_2 - k_1 + 1} \sum_{x_i=k_1}^{k_2} x_i \\
&= \frac{k_2(k_2+1)}{2(k_2 - k_1 + 1)} - \frac{k_1(k_1-1)}{2(k_2 - k_1 + 1)} \\
&= \frac{k_1 + k_2}{2}
\end{aligned}
$$

Likewise, the discrete uniform distribution's variance can be derived from;

$$
\begin{aligned}
Var[X] &= \sum_{x_i=k_1}^{k_2} (x_i - \mu_X)^2 \frac{1}{k_2 - k_1 + 1} = \sum_{x_i=k_1}^{k_2} \left(x_i - \frac{k_1 + k_2}{2}\right)^2 \frac{1}{k_2 - k_1 + 1} \\
&= \frac{(k_2 - k_1 + 1)^2 - 1}{12}
\end{aligned}
$$

Example 5.2 In a particular electrical system, the time required for a component to fail follows a discrete uniform distribution. The component can fail at any time between 36 and 72 months, inclusive. Let X be the random variable representing the time (in months) until the component fails.

(a) Derive the expected value (μ) of X, the time until the component fails.
(b) Calculate the variance (σ^2) of X.

Solution 5.2

(a) The expected value of a discrete uniform distribution is calculated as the average of all possible outcomes. For this electrical system, the minimum time (k_1) is 36 months, and the maximum time (k_2) is 72 months. The expected value is given by:

$$
\begin{aligned}
E[X] &= \frac{k_1 + k_2}{2} \\
&= \frac{36 + 72}{2} = 54
\end{aligned}
$$

(b) The variance of a discrete uniform distribution is given by the formula;

$$
\begin{aligned}
\sigma^2 &= \frac{(k_2 - k_1 + 1)^2 - 1}{12} \\
&= \frac{(72 - 36 + 1)^2 - 1}{12} = 114
\end{aligned}
$$

5.4.2 *Binomial Distribution*

The following conditions must be satisfied in order to have a proper binomial distribution;

1. The number of trials is fixed, i.e. n, for a specific random experiment.
2. Each trial is statistically identical and independent.

3. There are only two possible outcomes, such as "win" or "loose"; "successful" or "unsuccessful". They are usually notated as p, and q, where $p + q = 1$.
4. The probabilities of the two outcomes are constant during the entire experiment.

The probability density function of a binomial distribution is

$$f_X(x) = \sum_{k=0}^{n} \binom{n}{k} p^k (1-p)^{n-k} \delta(x-k) \tag{5.5}$$

Thereafter, the probability distribution function of a binomial distribution can be expressed as;

$$F_X(x) = \sum_{k=0}^{n} \binom{n}{k} p^k (1-p)^{n-k} u(x-k) \tag{5.6}$$

Hence, the corresponding expected value and the variance of a binomial distribution can be derived as;

$$E[X] = np$$
$$\text{Var}[X] = npq = np(1-p)$$

Some examples for the binomial distribution are illustrated in Fig. 5.3 for various n, and p values.

Please note that, Bernoulli distribution is a special version of binomial distribution where the number of trials, n, is fixed to 1.

Example 5.3 In a telecommunications network, data packets are transmitted through a channel. Assuming that 90% of packets are successfully transmitted without errors. If the system sends 50 packets, what is the probability that at least 45 of them are successfully transmitted without errors while packet transmissions are statistically independent.

Solution 5.3

The probability of k success out of n trials that can be calculated for a binomial distribution is given by:

$$P(X = k) = \binom{n}{k} p^k (1-p)^{n-k}$$

where: n is the number of trials (number of packets in this case), k is the number of successful trials (successfully transmitted packets), and p is the probability of success on a single trial.

In this example, $n = 50$, $k \geq 45$, and $p = 0.90$ (as 90% of packets are successfully transmitted without errors).

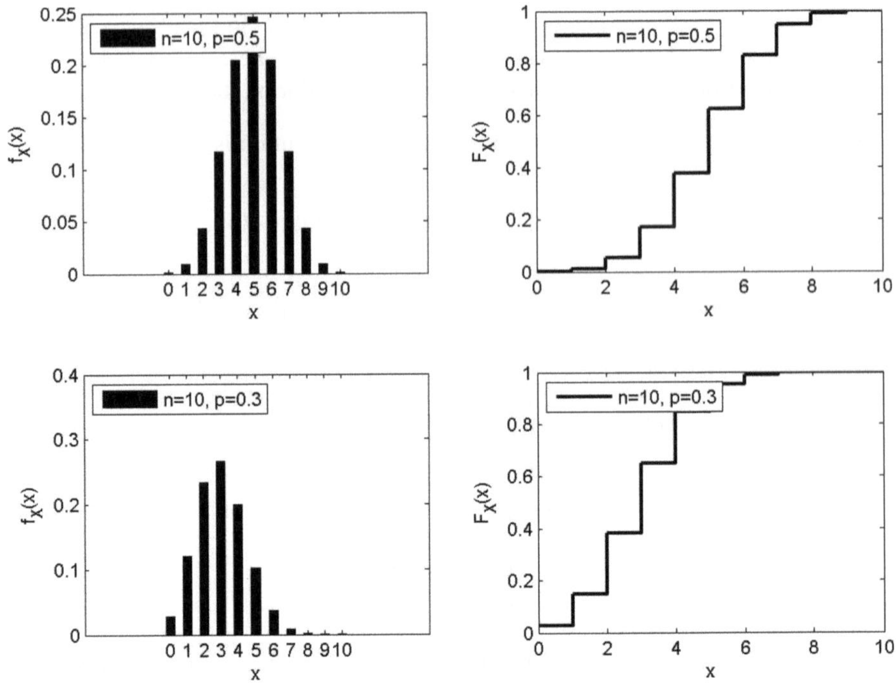

Fig. 5.3 Binomial probability distribution function, and probability density function

To find the probability of at least 45 successful transmissions;

$$P(X \geq 45) = P(X = 45) + P(X = 46) + P(X = 47) + P(X = 48) + P(X = 49) + P(X = 50)$$

$$P(X \geq 45) = P(X = 45) + P(X = 46) + P(X = 47) + P(X = 48) + P(X = 49) + P(X = 50)$$

$$\begin{aligned} P(X \geq 45) &= \binom{50}{45}(0.9)^{45}(0.1)^5 + \binom{50}{46}(0.9)^{46}(0.1)^4 + \binom{50}{47}(0.9)^{47}(0.1)^3 \\ &+ \binom{50}{48}(0.9)^{48}(0.1)^2 \binom{50}{49}(0.9)^{49}(0.1)^1 + \binom{50}{50}(0.9)^{50}(0.1)^0 \\ &= 0.6161 \end{aligned}$$

5.4.3 Pascal Distribution

The Pascal distribution, which is also known as the negative binomial distribution, describes the number of Bernoulli trials needed for a specified number of successes to take place. It is appropriate to employ the Pascal distribution in applications where the trials are independent and each trial may result in either "success" or "failure".

The Pascal distribution is characterized by three parameters, namely k, n and p, where k is the number of successes required, n is the number of trials needed to achieve k successes, and p is the probability of success on each independent trial (Fig. 5.4).

The probability density function of the Pascal distribution is given by:

$$f_X(x) = \sum_{n=k}^{\infty} \binom{n-1}{k-1} p^k (1-p)^{n-k} \delta(x-k) \tag{5.7}$$

The corresponding probability distribution function of the Pascal distribution is;

$$F_X(x) = \sum_{n=k}^{\infty} \binom{n-1}{k-1} p^k (1-p)^{n-k} u(x-k) \tag{5.8}$$

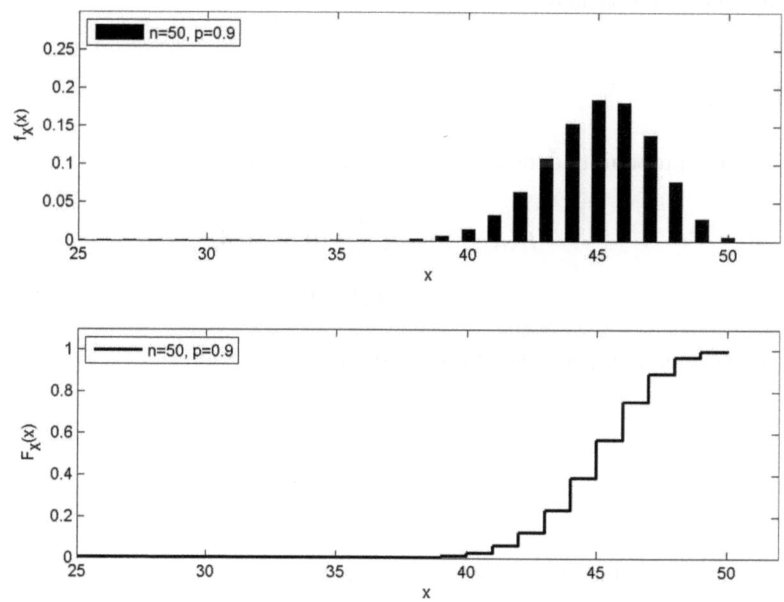

Fig. 5.4 The corresponding probability distribution function, and probability density function (Example 5.3)

The expected value and variance of the Pascal distribution are given by:

$$E[X] = \frac{k}{p}$$
$$\sigma^2 = \frac{k(1-p)}{p^2}$$

The Pascal distribution is often used in engineering applications such as analyzing the performance of a communication network; reliability analysis to model the time until a certain number of failures or faults occur in a system; and queuing theory involving the analysis of waiting times and the number of arrivals or departures until a certain event occurs.

Example 5.4 In a communication system, data packets are transmitted over a wireless channel. The transmission process is governed by a Pascal distribution with a probability of success on each transmission attempt $p = 0.6$

(a) Calculate the probability that it takes exactly 5 transmission attempts to successfully transmit 3 packets.
(b) Find the expected number of transmission attempts needed to successfully transmit 3 packets.
(c) Derive the variance of the number of transmission attempts needed to successfully transmit 3 packets.

Solution 5.4

(a) Using the probability density function of the Pascal distribution;

$$P(X = 5) = \binom{4}{2}(0.6)^3(0.4)^2$$
$$= 0.20736$$

(b) The expected number of attempts is calculated as;

$$E[X] = \mu_X = \frac{k}{p}$$
$$= \frac{3}{0.6} = 5$$

(c) The variance of the distribution is calculated as;

$$\sigma^2 = \frac{k(1-p)}{p^2}$$
$$= \frac{3(0.4)}{0.6^2} = 3.33$$

5.4.4 Poisson Distribution

The Poisson distribution describes the number of events that occur within a fixed interval of time. It is frequently employed to model rare events where the probability of occurrence is unlikely, and events happen independently of each other. The probability density function of a Poisson distribution can be given as;

$$f_X(x) = e^{-\lambda} \sum_{k=0}^{\infty} \frac{\lambda^k}{k!} \delta(x-k) \tag{5.9}$$

Hence, the corresponding probability distribution function is;

$$F_X(x) = e^{-\lambda} \sum_{k=0}^{\infty} \frac{\lambda^k}{k!} u(x-k) \tag{5.10}$$

The expected value and the variance of a Poisson distribution have the identical value of the parameter λ.

$$E[X] = \lambda$$
$$\mathrm{Var}[X] = \lambda$$

The Poisson distribution is memoryless, which means that previous events have no bearing on the likelihood of a particular number of events happening in the future.

Example 5.5 The Poisson distribution represents the incidence of errors in a telecommunications network, with an average rate of $\lambda = 0.5$ errors per minute.

(a) Find the probability that there will be exactly two errors in a minute.
(b) Calculate the expected number of errors in a one-minute interval.
(c) Assuming that the network is designed to tolerate up to 1 error per minute without significant impact on system performance. Calculate the probability that the network operates regularly in a given one-minute interval.

Solution 5.5

(a)

$$P(X=2) = f_X(2) = e^{-0.5} \frac{(0.5)^2}{2!}$$
$$= 0.0758$$

(b) The expected value of a Poisson distribution is λ, which is given as 0.5. $E[X] = 0.5$
(c) The condition that the network operates regularly in a given one-minute interval, $P(X \leq 1)$, therefore

$$\begin{aligned} P(X \leq 1) &= P(X=0) + P(X=1) \\ &= e^{-0.5}\frac{(0.5)^0}{0!} + e^{-0.5}\frac{(0.5)^1}{1!} \\ &= 0.6065 + 0.3033 \\ &= 0.9098 \end{aligned}$$

It can be stated that, the network system operates regularly with almost 91% of probability (Table 5.1).

5.5 Characteristic Function

The characteristic function of a discrete random variable X is defined as follows;

$$\Phi_X(\omega) = E\left[e^{j\omega x}\right] = \sum_{x \in R_X} e^{jwx} f_X(x) \tag{5.11}$$

It is directly related the Discrete Time Fourier Transform (DTFT) of the probability density function $f_X(x)$ with the negative sign of ω. One of the main advantages of the characteristic function, there exists always a $\Phi_X(\omega)$ of a proper probability distribution. In addition, considering the statistical analysis of a random variable, the nth moment of the distribution, that is m_n, can be easily calculated by means of derivative operator;

$$m_n = \lim_{w \to 0} (-j)^n \frac{\mathrm{d}^n \Phi_X(\omega)}{\mathrm{d}w^n} \tag{5.12}$$

5.5.1 Properties of Characteristic Function

Some of the significant properties of a characteristic functions can be listed as;

The characteristic function, $\Phi_X(w)$, exists for any probability distribution of the random variable X.
Since the sum of all possibilities must be 1, then $\Phi_X(0) = 1$.
The characteristic function has its maximum value at $\omega = 0$; therefore, $|\Phi_X(\omega)| \leq \Phi_X(0) = 1$ for all ω
The characteristic function of $Y = k_1 X + k_2$ is equal to $\Phi_Y(\omega) = e^{jk_2\omega}$ $\Phi_X(k_1\omega)$

Table 5.1 The probabilities of $P(X \leq x)$ governing by a Poisson distribution at various λ, and x values

λ/x	0	1	2	3	4	5	6	7	8	9	10
1.0	0.3679	0.7358	0.9197	0.9817	0.9966	0.9996	0.9999	1.0000	1.0000	1.0000	1.0000
1.5	0.1889	0.4225	0.6767	0.8544	0.9473	0.9839	0.9949	0.9986	0.9996	0.9999	1.0000
2.0	0.1353	0.4060	0.6767	0.8571	0.9473	0.9820	0.9939	0.9982	0.9996	0.9999	1.0000
2.5	0.1032	0.3447	0.6038	0.8194	0.9319	0.9758	0.9912	0.9970	0.9990	0.9997	0.9999
3.0	0.0498	0.1495	0.2990	0.5034	0.6988	0.8361	0.9161	0.9578	0.9789	0.9899	0.9954
3.5	0.0286	0.1074	0.2388	0.4111	0.6007	0.7580	0.8728	0.9427	0.9764	0.9919	0.9970
4.0	0.0183	0.0731	0.1830	0.3670	0.6109	0.8158	0.9165	0.9676	0.9886	0.9966	0.9988
4.5	0.0111	0.0501	0.1282	0.2682	0.4512	0.6268	0.7695	0.8711	0.9375	0.9748	0.9903
5.0	0.0067	0.0335	0.1006	0.2003	0.3344	0.4405	0.5353	0.6146	0.6827	0.7420	0.7946
5.5	0.0041	0.0227	0.0687	0.1446	0.2423	0.3322	0.4115	0.4762	0.5307	0.5771	0.6179
6.0	0.0025	0.0148	0.0444	0.0888	0.1480	0.2035	0.2562	0.3082	0.3605	0.4132	0.4668
6.5	0.0015	0.0092	0.0275	0.0551	0.0914	0.1262	0.1590	0.1899	0.2188	0.2457	0.2709
7.0	0.0009	0.0063	0.0188	0.0376	0.0628	0.0904	0.1164	0.1394	0.1596	0.1779	0.1947
7.5	0.0005	0.0039	0.0115	0.0230	0.0383	0.0551	0.0714	0.0857	0.0985	0.1099	0.1202
8.0	0.0003	0.0021	0.0063	0.0126	0.0210	0.0301	0.0390	0.0468	0.0537	0.0599	0.0654
8.5	0.0002	0.0013	0.0039	0.0078	0.0130	0.0187	0.0241	0.0290	0.0333	0.0372	0.0407
9.0	0.0001	0.0007	0.0022	0.0045	0.0075	0.0108	0.0140	0.0169	0.0193	0.0212	0.0229
9.5	0.0000	0.0004	0.0013	0.0027	0.0045	0.0065	0.0085	0.0102	0.0118	0.0131	0.0143
10.0	0.0000	0.0003	0.0010	0.0020	0.0034	0.0048	0.0061	0.0073	0.0085	0.0096	0.0106

Table 5.2 Some of discrete probability distributions and their parameters

Distribution	$f_X(x)$	$E[X] = \mu_X$	$\text{Var}[X]$	$\Phi_X(w)$
Uniform	$\sum_{k=k_1}^{k_2} \frac{1}{k_2-k_1+1}\delta(x-k)$	$\frac{k_1+k_2}{2}$	$\frac{(k_2-k_1+1)^2-1}{12}$	$\frac{e^{j\omega k_1}-e^{j\omega(k_2+1)}}{(1-e^{j\omega})(k_2-k_1+1)}$
Binomial	$\sum_{k=0}^{n} \binom{n}{k} p^k(1-p)^{n-k}\delta(x-k)$	np	$np(1-p)$	$(1-p+pe^{j\omega})^n$
Pascal	$\sum_{n=k}^{\infty} \binom{n-1}{k-1} p^k(1-p)^{n-k}\delta(x-k)$	$\frac{k}{p}$	$\frac{k(1-p)}{p^2}$	$p^k e^{j\omega k}\left(1-(1-p)e^{j\omega}\right)^{-k}$
Poisson	$e^{-\lambda}\sum_{k=0}^{\infty} \frac{\lambda^k}{k!}\delta(x-k)$	λ	λ	$e^{\lambda(e^{j\omega}-1)}$

The characteristic function of $Y = -X$ is the complex conjugate $\Phi_X^*(\omega)$, i.e. $\Phi_Y(\omega) = \Phi_X^*(\omega)$
Let Z be a random variable fined as convolution of two random variables such that $Z = X * Y$, then it can be stated that

$$\Phi_Z(\omega) = \Phi_X(\omega)\Phi_Y(\omega)$$

Example 5.6 Derive the characteristic function of the Poisson distribution (Table 5.2).

Solution 5.6

$$\begin{aligned}
\Phi_X(\omega) &= \sum_{x=0}^{\infty} e^{j\omega x}\left(\sum_{k=0}^{\infty} e^{-\lambda}\frac{\lambda^k}{k!}\delta(x-k)\right) \\
&= \sum_{x=0}^{\infty} e^{j\omega x} e^{-\lambda}\frac{\lambda^x}{x!} \\
&= e^{-\lambda}\sum_{x=0}^{\infty} \frac{\left(\lambda e^{j\omega}\right)^x}{x!} \\
&= e^{-\lambda} e^{\lambda e^{j\omega}} \\
&= e^{\lambda(e^{j\omega}-1)}
\end{aligned}$$

Problems

5.1 Consider a cryptographic system that generates random keys, and it uses a discrete uniform distribution to select a key from the set $\{0, 1, 2, \ldots, 255\}$. Let X be the random variable representing the selected key.

(a) Calculate the expected value (μ) and variance (σ^2) of the random variable X.
(b) Find the probability that the selected key is divisible by 5.

5.2 In a parallel computing environment, tasks are distributed among m processors, each assigned an ID from the set $\{1, 2, \ldots, m\}$ using a discrete uniform distribution. Let Y be the random variable representing the ID of a randomly selected processor.

(a) Determine the probability that a specific processor with ID 3 is selected when $m = 32$
(b) Derive the expected value (μ) and variance (σ^2) of the random variable Y representing the selected processor ID.

5.3 A binomial distribution is frequently used to model a cyberattack's success in network security. Let X be the random variable with a success probability of $p = 0.05$, representing the number of successful infiltrations.

(a) Determine the probability that, out of 100 hacking attempts, only 4 will be successful.
(b) Calculate the probability that, out of 100 hacking attempts, only 4 will be successful, under the condition that the first 50 attempts were unsuccessful.

5.4 A Pascal distribution with a success probability of $p = 0.45$ governs the number of attempts in a telecommunications network until 3 successful connections get established.

(a) Calculate the probability that exactly 8 attempts are needed to establish 3 successful connections.
(b) Find the mean (μ) and standard deviation (σ) of the number of attempts until 3 successful connections are established.

5.5 In a complex system, the time in seconds until 5 critical failures occur follows a Pascal distribution with a failure probability $p = 0.01$

(a) Find the expected value (μ) and variance (σ^2) of the time until 5 critical failures occur.
(b) Determine the probability that exactly 20 s are needed to experience 2 critical failures.
(c) Calculate the probability that at most 8 s are needed to experience 5 critical failures. Discuss about the reliability of the system.

5.6 The number of data packets in a computer network arriving at a router per second follows a Poisson distribution with an average rate of $\lambda = 3$

(a) Determine the expected number of packets arriving in a minute.
(b) Calculate the probability that exactly 120 packets arrive in a minute.
(c) If the network is designed to handle up to 240 packets per minute without congestion, find the probability that the network operates properly.
(d) How an increase in the average arrival rate (λ) might affect the network's performance and the probability of congestion?

5.7 A computer system experiences failures. The number of failures per day follows a Poisson distribution with a mean of $\lambda = 0.6$.

(a) What is the probability of having exactly 2 failures in 3 days?
(b) What is the probability of having at least 1 failure in a day?

5.8 In a communication channel, the number of bit errors per transmission is modeled as a binomial random variable. If the probability of a bit error is $p = 10^{-4}$, and 10^6 bits are transmitted:

(a) What is the probability of exactly 100 bit errors?
(b) Approximate the probability of at most 5 bit errors using the Poisson distribution.

5.9 The number of users accessing a server simultaneously is a random variable X with a discrete uniform distribution over the integers from 10 to 50 (inclusive).

(a) What is the probability that exactly 25 users are accessing the server?
(b) What is the expected number of users accessing the server?

5.10 The time it takes to crack a password, in hours, follows a Pascal distribution. If the probability of cracking a password in any given attempt is $p = 0.2$, and we want to model the time until the 5th successful crack:

(a) What is the probability that it takes exactly 20 h to crack 5 passwords? (Assume each attempt takes 1 h.)
(b) What is the expected time to crack 5 passwords?

5.11 In a network, packets are routed through a series of nodes. The probability that a packet is dropped at any node is $p = 0.01$. Consider sending 1000 packets.

(a) What is the probability that exactly 15 packets are dropped?
(b) What is the probability that at least 5 packets are dropped?

5.12 The number of requests received by a web server per minute follows a Poisson distribution with a mean of $\lambda = 5$.

(a) What is the probability that exactly 10 requests are received in 2 min?
(b) What is the probability that no requests are received in 30 s?

5.13 A hard drive manufacturer finds that 2

(a) What is the probability that exactly 3 drives fail within the first year?
(b) What is the probability that more than 5 drives fail within the first year?

5.14 In a local area network, the number of packets colliding in a 10 ms interval is a Poisson random variable with mean 0.8.

(a) What is the probability of no collisions in a 25 ms interval?
(b) What is the probability of at least 1 collision in a 5 ms interval?

5.15 A security system logs events. The number of events logged in an hour is uniformly distributed between 0 and 10.

(a) What is the probability that exactly 5 events are logged in an hour?
(b) What is the average number of events logged in 3 h?

5.16 The number of attempts to establish a connection in a network until the 4th successful connection is achieved follows a Pascal distribution. The probability of a successful connection is 0.8.

(a) What is the probability that it takes exactly 6 attempts to achieve 4 successful connections?
(b) What is the probability that it takes at most 5 attempts to achieve 4 successful connections?

5.17 In a data center, the probability of a server failure in a day is 0.001. There are 1000 servers.

(a) What is the probability that exactly 2 servers fail in a day?
(b) What is the probability that at least one server fails in a day?

5.18 The number of print jobs arriving at a network printer per minute is a Poisson random variable with a mean of 4.

(a) What is the probability that exactly 5 print jobs arrive in a minute?
(b) What is the probability that fewer than 3 print jobs arrive in 30 s?

5.19 A system logs errors. The number of errors logged per hour is uniformly distributed between 0 and 15.

(a) What is the probability that exactly 10 errors are logged in an hour?
(b) What is the probability that between 5 and 10 errors (inclusive) are logged in an hour?

5.20 The number of attempts needed to send a message successfully across a noisy channel follows a Pascal distribution. The probability of successful transmission is 0.7. We are interested in the number of attempts until the 3rd successful transmission.

(a) What is the probability that it takes exactly 5 attempts?
(b) What is the expected number of attempts needed?

Chapter 6
Random Processes

A system may generally be random in its nature; it may also be deterministic with an unwanted random noise feature; as an alternative such a process may consist of both a random main component and a random noisy part. In each of these scenarios, the system's uncertainty and the time aspect should to be considered during analysis. In this chapter, discrete time random processes will be discussed in detail. The basic definition of a discrete time random process is a sequence of random variables indexed with time in most applications [1]. It might be seen a realization of random variables over a certain time interval. Therefore, the probability theory is the basis of random processes by means of random variables.

In Chap. 5, we introduced the concept of a *random variable* X, which is a function that maps each possible outcome s to a value in the real number space $\mathbb{R}$. Now, we extend this idea further by defining a function that depends not only on the possible outcomes s but also on discrete in time n. This function, denoted as $X(s, n)$, is referred to as a *random process*.

To better understand the structure of a random process, we consider the following notational conventions:

Ensemble Representation: The collection of all possible functions $X(s, n)$ for different values of s is called an *ensemble*. It represents the entire set of possible outcomes of the random process over time.

Realization (Sample Function): If we fix s at a specific value, say s_1, then the function $X(s_1, n)$ becomes a deterministic function of time n. This specific instance of the process is known as a *realization* or a *sample function*.

Time Sample (Random Variable at a Fixed Time): If we hold time n constant at a particular value n_1, then the function $X(s, n_1)$ reduces to a *random variable*, as it depends only on the random outcome s. This is referred to as a *time sample*.

M. Catak and T. Allahviranloo, *Discrete Random Processes*, Studies in Systems, Decision and Control 644, https://doi.org/10.1007/978-3-032-08630-3_6

Single Numerical Value: When both s and n are fixed at specific values, the function $X(s, n)$ evaluates to a single numerical value, losing its stochastic nature.

Let define a random variable X be uniformly distributed in the interval of $[1, 10]$ 20 realization have been generated and each realization contains 30 random experiments. The results are given in Table 6.1.

Ensemble is the family function that may contain all possible realization of a random process. In Table 6.1 30 realizations has been carried out as illustrative examples. Each row of the Table represents a random process; while each column represents a random variable (since the time index is fixed at a certain column). The mean and the standard deviation of each realization are also given in the Table. Since the governing random variable is defined as a uniform distribution for this example, its mean and the standard deviation can be easily calculated as;

$$\mu_X = \frac{1+10}{2} = 5.5$$

$$\sigma_X = \sqrt{\frac{(10-1+1)^2-1}{12}} = 2.8723$$

The average parameters obtained from the realizations of a random process are usually different from their theoretical values. It will be discussed under the next sections. In this textbook, the following notations will be adopted;

- μ_X, and $E[X]$:= The mean (or the expected value) of the random variable govern by a proper probability distribution function.
- σ_X := The standard deviation of the random variable govern by a proper probability distribution function.
- $\mu_X(n)$, and $E[X(n)]$:= The mean (or the expected value) of the realization (the generated random sequence) of a random signal.
- $\sigma_X(n)$:= The standard deviation of the realization (the generated random sequence) of a random signal.

6.1 Averages

In probability and statistics, one of the fundamental ways to describe a random variable is through its statistical averages, such as the mean and variance. These measures provide insights into the central tendency and dispersion of the variable, respectively. However, when dealing with random processes, which consist of a

Table 6.1 Realization of a random process based on uniformly distributed integers in the interval of [1, 10]

Ensemble	Sequence of a random sample (time index n)																														Mean	Std
	1	2	3	4	5	6	7	8	9	10	11	12	13	14	15	16	17	18	19	20	21	22	23	24	25	26	27	28	29	30		
1	8	9	5	8	8	4	2	6	10	2	6	3	8	9	3	6	3	4	1	8	1	5	3	9	1	7	5	3	6	7	5.33	2.72
2	10	4	3	2	9	1	7	9	6	7	4	7	9	6	4	5	2	10	6	5	6	5	9	7	7	8	10	4	3	10	6.17	2.64
3	6	5	10	4	2	8	6	10	8	4	1	2	8	8	3	3	4	6	3	8	8	6	9	5	3	4	3	8	3	3	5.37	2.59
4	3	9	2	4	1	10	8	9	5	4	4	6	6	5	9	6	8	8	5	10	4	1	3	3	3	8	7	9	7	8	5.83	2.69
5	3	8	9	9	6	2	2	3	9	6	3	8	8	9	10	9	8	9	7	6	5	7	7	3	4	9	10	4	9	9	6.70	2.59
6	8	3	1	1	7	2	4	1	4	9	3	10	5	4	7	1	8	2	3	4	3	1	8	5	2	1	7	10	9	5	4.60	2.99
7	10	8	4	9	6	10	2	5	8	2	9	8	9	10	8	5	1	10	7	8	10	6	10	3	6	2	3	10	3	2	6.47	3.08
8	4	3	8	3	1	5	9	1	10	7	4	8	1	3	10	1	4	2	4	3	5	8	10	1	10	9	6	3	4	9	5.20	3.18
9	8	3	7	3	7	9	4	5	6	10	1	5	10	1	7	6	10	8	9	3	6	5	4	2	4	5	1	8	9	5	5.70	2.77
10	2	1	2	5	10	9	3	10	5	2	4	3	1	8	6	6	4	8	2	5	3	7	2	3	6	5	9	7	3	2	4.77	2.74
11	6	4	10	8	9	2	1	3	3	3	8	6	6	9	4	8	6	8	2	7	1	3	8	2	7	10	5	10	10	9	5.93	2.97
12	5	5	6	3	2	3	9	4	4	5	2	3	5	10	7	3	3	2	10	7	1	7	2	7	8	8	4	7	7	7	5.20	2.54
13	8	4	3	3	4	1	5	5	2	9	10	9	2	4	4	6	2	5	2	2	9	4	2	9	5	10	10	8	2	1	5.00	3.03
14	3	3	6	10	3	7	6	9	4	6	4	7	8	9	5	5	4	6	2	2	8	2	4	9	9	3	5	8	9	10	5.87	2.58
15	1	3	10	6	6	8	9	9	4	3	4	9	7	6	4	9	9	6	2	3	2	4	9	9	10	1	3	4	7	1	5.60	2.99
16	5	10	9	4	7	6	8	5	4	6	3	8	8	5	5	7	10	9	1	8	9	3	6	8	8	4	1	3	6	1	5.90	2.64
17	7	9	5	10	4	1	7	4	9	6	8	5	6	9	6	4	9	7	5	5	7	3	6	9	3	6	6	1	3	10	6.00	2.48
18	2	4	5	9	8	10	2	10	6	5	1	4	9	10	10	4	1	9	7	9	7	7	10	8	8	6	2	5	1	7	6.20	3.04
19	5	5	4	2	8	6	6	8	6	1	10	7	6	1	3	3	4	6	9	10	10	6	6	3	2	10	1	7	3	4	5.40	2.81
20	7	6	1	1	8	4	10	8	8	8	2	10	10	8	4	10	3	10	8	4	1	5	1	10	6	5	6	3	4	6	5.90	3.07

collection of indexed random variables evolving over time or another parameter, it becomes essential to extend these statistical descriptions to the entire process.

To systematically analyze a random process, we rely on ensemble averages; statistical measures computed across all possible realizations (or outcomes) of the process. These averages help in understanding the overall behavior of the process rather than just individual realizations.

Before going into the specifics of how these averages are computed and utilized, it is important to establish some fundamental definitions that will serve as the basis for further discussion.

Expected Value, $\mu_X(n)$

The expected value of a discrete random process $X(n)$ is given by:

$$E[X(n)] = \mu_X(n) = \sum_{\forall x} x_i f_{X(n)}(x_i) \tag{6.1}$$

where x_i represents the possible values of $X(n)$, and $f_{X(n)}$ is the corresponding probability density function.

Variance $Var[X(n)]$, and Standard Deviation $\sigma_{X(n)}$

In the analysis of a discrete random process, the variance and the standard deviation of a random process provide insight into the spread and dispersion of the random process around its mean. The variance of a discrete random process $X(n)$ quantifies how much the values of the process deviate from the expected value (mean). It is defined as:

$$Var[X(n)] = E\left[(X(n) - E[X(n)])^2\right] = \sum_i (x_i - E[X(n)])^2 f_{X(n)}(x_i) \tag{6.2}$$

The standard deviation is simply the square root of the variance and provides a measure of the typical deviation from the mean:

$$\sigma_{X(n)} = \sqrt{Var[X(n)]}$$

A lower standard deviation indicates that the values of $X(n)$ are closely clustered around the mean, while a higher standard deviation signifies greater dispersion.

6.2 Correlation Functions

In general, the first-order statistical characteristics of a random process, such as its mean and variance, provide only limited insights into the underlying nature of the process. These measures do not capture the intricate dependencies and structural properties of the signal. To gain deeper insight into the signal characteristics, the autocorrelation function is frequently employed, as it facilitates the analysis of the power distribution within the signal. Mathematically, the autocorrelation function is expressed as:

$$\begin{aligned} R_{XX}[m,n] &= \sum_{m=-\infty}^{\infty}\sum_{n=-\infty}^{\infty} x[m]x^*[n] \\ R_{XX}[k] &= \sum_{m=-\infty}^{\infty} x[m]x^*[m-k] \end{aligned} \tag{6.3}$$

where $k = m - n$, and $*$ denotes the complex conjugate of the random signal. The autocorrelation function essentially quantifies the degree of similarity between a signal and its time-shifted version, thereby offering valuable information about the temporal structure and periodicity of the process.

Beyond analyzing a single random signal, it is often crucial to evaluate the relationship between two different random signals. This can be achieved through the cross-correlation function, which measures the statistical dependence between two signals. The cross-correlation function is defined as:

$$\begin{aligned} R_{XY}[m,n] &= \sum_{m=-\infty}^{\infty}\sum_{n=-\infty}^{\infty} x[m]y^*[n] \\ R_{XY}[k] &= \sum_{m=-\infty}^{\infty} x[m]y^*[m-k] \end{aligned} \tag{6.4}$$

The cross-correlation function serves as an effective tool for quantifying the similarity between two random signals, allowing for the detection of relationships and dependencies between them. This function finds widespread applications across various engineering fields, including signal processing, wireless communications, and system identification.

To facilitate meaningful comparisons across different scenarios, the correlation value is often normalized, which is called correlation coefficient. The raw correlation values depend on the absolute magnitude of the signals, which makes direct comparisons difficult when dealing with signals of varying power levels or amplitudes. Normalization ensures that correlation values remain within a fixed range, typically between –1 and 1, allowing for a consistent and scale-independent interpretation of similarity. The correlation coefficient can be expressed as;

$$\rho_{XY}[k] = \frac{R_{XY}[k]}{\sqrt{R_{XX}[0]R_{YY}[0]}} \tag{6.5}$$

where $R_{XX}[0]$ and $R_{YY}[0]$ represent the autocorrelation values of signals $x[m]$ and $y[m]$ at zero lag, corresponding to their total power. The normalization factor ensures that the correlation coefficient remains bounded within:

$$-1 \leq \rho_{XY}[k] \leq 1. \tag{6.6}$$

Normalization plays a crucial role in correlation analysis for several reasons. First, it ensures "scale independence", meaning that the correlation measure remains unaffected by differences in signal amplitude. This is particularly important when comparing signals with different power levels. Second, it improves "interpretability", as the normalized correlation coefficient provides a clear and standardized measure of similarity. A value of $\rho_{XY}[k] = 1$ indicates a perfect positive correlation, implying that the two signals are identical up to a scaling factor. Conversely, $\rho_{XY}[k] = -1$ signifies a perfect negative correlation, meaning that the signals are mirror images of each other. A correlation coefficient of zero, $\rho_{XY}[k] = 0$, suggests no linear relationship between the signals. Lastly, normalization is essential for applications in fields such as signal processing, wireless communications, pattern recognition, and system identification, where detecting relationships between signals is crucial, regardless of their absolute power levels.

By employing normalization, the correlation function becomes a robust and reliable metric for evaluating signal similarity, making it a valuable tool in engineering applications.

6.3 Covariance Functions

In signal processing and statistical analysis, correlation and covariance functions are fundamental tools used to quantify the relationship between random signals. While both functions measure dependency, they have different characteristics and applications.

The correlation function measures the overall similarity between two signals, capturing linear dependency and periodicity. It does not explicitly remove the mean values of the signals, making it useful for analyzing relationships in various signal processing applications. The covariance function, in contrast, measures how two signals vary together relative to their respective mean values. Unlike correlation, covariance explicitly subtracts the mean from each signal before computing their joint variability. The autocovariance function is defined as;

$$\begin{aligned} C_{XX}[m,n] &= E\left[(x[m]-\mu_X(m))\,(x[n]-\mu_X(n))^*\right] \\ C_{XX}[m,n] &= R_{XX}[m,n]-\mu_X(m)\mu_X^*(n) \end{aligned} \tag{6.7}$$

Table 6.2 Comparison of the covariance and correlation functions

Feature	Correlation function $R_{XY}[k]$	Covariance Function $C_{XY}[k]$
Definition	Measures overall similarity, including periodicity	Measures joint variability around the mean
Mean subtraction	Mean is not subtracted	Mean is subtracted before computation
Scale dependence	Can be normalized for scale invariance	Affected by signal magnitudes
Range	Can be normalized to $[-1, 1]$	Unbounded
Application	Signal processing, pattern recognition	Statistical analysis, feature extraction

Let define the time lag $k = m - n$, then the autocovariance function can be explained;

$$C_{XX}[k] = R_{XX}[k] - \mu_X(m)\mu_X^*(m-k) \tag{6.8}$$

In the same manner, the crosscovariance function between two random signal is derived;

$$C_{XY}[k] = R_{XY}[k] - \mu_X(m)\mu_Y^*(m-k) \tag{6.9}$$

The key differences between the covariance and correlation functions are summarized in Table 6.2.

Both correlation and covariance functions are essential in analyzing dependencies between signals, but they serve different purposes. The correlation function provides a general measure of similarity, making it widely used in signal processing, pattern recognition, and system identification. On the other hand, the covariance function focuses on deviations from the mean, making it more suitable for statistical analysis and feature extraction. The normalization of correlation further enhances its applicability by providing a standardized, scale-independent measure of similarity.

6.4 Properties of Correlation and Covariance Sequences

The correlation and covariance functions exhibit several fundamental properties that provide insights into the characteristics of random signals and their relationships. These properties are essential in statistical signal processing and stochastic analysis.

Symmetry Property

The autocorrelation function exhibits Hermitian symmetry:

$$R_{XX}[k] = R^*_{XX}[-k] \tag{6.10}$$

This property signifies that the autocorrelation sequence is conjugate symmetric. Consequently, the real part of the autocorrelation function is an even function, $Re\{R_{XX}[k]\} = Re\{R_{XX}[-k]\}$, and the imaginary part is an odd function, $Im\{R_{XX}[k]\} = -Im\{R_{XX}[-k]\}$. For real-valued signals, this reduces to even symmetry, $R_{XX}[k] = R_{XX}[-k]$. The Fourier transform of a Hermitian symmetric sequence is always real-valued, a crucial aspect in spectral analysis.

The autocovariance function of a wide-sense stationary process is an even function:

$$C_{XX}[k] = C_{XX}[-k] \tag{6.11}$$

For a wide-sense stationary (WSS) discrete-time random process $x[n]$ with a constant mean μ_x, the autocovariance function $C_{XX}[k]$, defined as $E\{(x[n] - \mu_x)(x[n-k] - \mu_x)^*\}$, satisfies this even symmetry property. This indicates that the covariance between the process at a given time and at a time lagged by k is identical to the covariance between the process at that time and at a time advanced by k. While the autocorrelation of a complex WSS process is Hermitian symmetric, the removal of the constant mean in the autocovariance calculation results in this even symmetric behavior.

The cross-correlation function demonstrates conjugate relationship under lag reversal:

$$R_{XY}[k] = R^*_{YX}[-k] \tag{6.12}$$

For two discrete-time signals $x[n]$ and $y[n]$, the cross-correlation functions $R_{XY}[k] = E\{x[n]y^*[n-k]\}$ and $R_{YX}[k] = E\{y[n]x^*[n-k]\}$ are related by this expression. This relationship implies that the cross-correlation between the signal x at time n and the signal y at a time lagged by k is the complex conjugate of the cross-correlation between the signal y at time m and the signal x at a time lagged by $-k$ (or advanced by k, where $m = n - k$). For real-valued signals, this simplifies to $R_{XY}[k] = R_{YX}[-k]$, indicating a direct even symmetry upon reversal of the roles of the two signals and the lag. This property is fundamental in analyzing the temporal relationships between different signals.

Maximum at Zero Lag

The maximum value of the autocorrelation function is attained at zero lag:

$$|R_{XX}[k]| \leq R_{XX}[0] \tag{6.13}$$

where $R_{XX}[0] = E\{|x[n]|^2\}$ represents the average power of the discrete-time signal $x[n]$. This property arises from the fact that a signal is maximally correlated with itself. At any non-zero lag $k \neq 0$, the correlation will be less than or equal to the power of the signal at zero lag. For real-valued signals, this simplifies to $-R_{XX}[0] \leq R_{XX}[k] \leq R_{XX}[0]$. The magnitude of the autocorrelation function provides a measure of the self-similarity of the signal as a function of the time lag, with the peak at $k = 0$ indicating perfect alignment.

The cross-correlation function is bounded by the geometric mean of the individual signal powers, satisfying the Cauchy-Schwarz inequality:

$$|R_{XY}[k]|^2 \leq R_{XX}[0]R_{YY}[0] \tag{6.14}$$

where $R_{XX}[0] = E\{|x[n]|^2\}$ and $R_{YY}[0] = E\{|y[n]|^2\}$ represent the average powers of the discrete-time signals $x[n]$ and $y[n]$, respectively. This inequality establishes an upper bound on the magnitude squared of the cross-correlation function between two signals at any lag k. It is a direct consequence of the Cauchy-Schwarz inequality applied to the inner product space of random variables. The equality $|R_{XY}[k]|^2 = R_{XX}[0]R_{YY}[0]$ holds if and only if the signals $x[n]$ and $y[n-k]$ are linearly dependent. This bound is crucial in various signal processing applications, including detection and estimation theory, as it provides a fundamental limit on the degree of linear relationship that can exist between two signals.

Relationship Between Correlation and Covariance

The autocovariance function is expressed in terms of the autocorrelation function and the squared mean of the signal:

$$C_{XX}[k] = R_{XX}[k] - \mu_X \mu_X^* \tag{6.15}$$

where $C_{XX}[k] = E\{(x[n] - \mu_X)(x[n-k] - \mu_X)^*\}$ is the autocovariance function and $R_{XX}[k] = E\{x[n]x^*[n-k]\}$ is the autocorrelation function of the discrete-time signal $x[n]$ with mean $\mu_X = E\{x[n]\}$. This relationship explicitly demonstrates that the autocovariance quantifies the correlation between deviations of the signal from its mean at different time instances, whereas the autocorrelation measures the correlation between the signal values themselves at those instances. For signals with a zero mean ($\mu_X = 0$), the autocovariance function is identical to the autocorrelation function.

The cross-covariance function is defined in relation to the cross-correlation function and the product of the individual signal means:

$$C_{XY}[k] = R_{XY}[k] - \mu_X \mu_Y^* \tag{6.16}$$

where $C_{XY}[k] = E\{(x[n] - \mu_X)(y[n-k] - \mu_Y)^*\}$ is the cross-covariance function and $R_{XY}[k] = E\{x[n]y^*[n-k]\}$ is the cross-correlation function between two discrete-time signals $x[n]$ and $y[n]$ with means $\mu_X = E\{x[n]\}$ and $\mu_Y = E\{y[n]\}$, respectively. The cross-covariance thus measures the correlation between the deviations of the two signals from their respective means at different time instances. If either or both signals have a zero mean ($\mu_X = 0$ or $\mu_Y = 0$), the cross-covariance function simplifies to the cross-correlation function. This distinction is crucial when analyzing the linear dependence between signals while accounting for their average values.

Relation to Power Spectral Density

The Power Spectral Density (PSD) is obtained by taking the Fourier Transform of the autocorrelation function:

$$S_{XX}(e^{j\omega}) = \sum_{k=-\infty}^{\infty} R_{XX}[k]e^{-j\omega k} \tag{6.17}$$

Similarly, the cross-power spectral density is:

$$S_{XY}(e^{j\omega}) = \sum_{k=-\infty}^{\infty} R_{XY}[k]e^{-j\omega k} \tag{6.18}$$

Linearity Property

Correlation satisfies linearity:

$$R_{\alpha X+\beta Y,Z}[k] = \alpha R_{XZ}[k] + \beta R_{YZ}[k] \tag{6.19}$$

where α, β are constants.

Normalized Correlation

The normalized autocorrelation function is defined as:

$$\rho_{XX}[k] = \frac{R_{XX}[k]}{\sqrt{R_{XX}[0]R_{XX}[0]}} \tag{6.20}$$

For cross-correlation:

$$\rho_{XY}[k] = \frac{R_{XY}[k]}{\sqrt{R_{XX}[0]R_{YY}[0]}} \tag{6.21}$$

ensuring that:

$$-1 \leq \rho_{XY}[k] \leq 1 \tag{6.22}$$

Ergodicity and Time Averaging

For an ergodic process, the autocorrelation function can be equivalently determined through a temporal average:

$$R_{XX}[k] = \lim_{N\to\infty} \frac{1}{N} \sum_{m=0}^{N-1} x[m]x^*[m-k] \tag{6.23}$$

Ergodicity, in the context of random processes, implies that the statistical properties of the process can be inferred from a single sufficiently long realization (time series) of the process. Consequently, the ensemble average, which defines the autocorrelation function $R_{XX}[k] = E\{x[n]x^*[n-k]\}$, can be replaced by the time average shown above. This property is of significant practical importance as it allows for the estimation of the autocorrelation function directly from observed data without requiring knowledge of the underlying probabilistic distribution or multiple realizations of the process. The convergence of this time average to the ensemble average is a fundamental characteristic of ergodic processes, enabling statistical inference from temporal observations.

To summarize, the key properties of correlation and covariance functions are outlined in Table 6.3.

Table 6.3 Summary of correlation and covariance Properties

Property	Description
Symmetry	$R_{XX}[k] = R^*_{XX}[-k]$, $R_{XY}[k] = R^*_{YX}[-k]$
Maximum at zero lag	$\lvert R_{XX}[k]\rvert \le R_{XX}[0]$
Covariance relation	$C_{XY}[k] = R_{XY}[k] - \mu_X \mu_Y$
Spectral relation	$S_{XX}(e^{j\omega}) = \mathcal{F}\{R_{XX}[k]\}$
Linearity	$R_{\alpha X+\beta Y,Z}[k] = \alpha R_{XZ}[k] + \beta R_{YZ}[k]$
Normalization	$-1 \le \rho_{XY}[k] \le 1$
Ergodicity	$R_{XX}[k] = \lim_{N\to\infty} \frac{1}{N} \sum_{m=0}^{N-1} x[m]x^*[m-k]$

6.5 Fourier Transform in Discrete-Time Signal Processing

In discrete-time signal processing, analyzing signals in the time domain alone is often insufficient to fully understand their characteristics and behavior. Many real-world signals, such as speech, audio, biomedical signals, and communication waveforms, contain complicated frequency components that are not immediately apparent in the time domain. The Fourier Transform provides a fundamental tool to bridge this gap by converting discrete-time signals from the time domain to the frequency domain, allowing for a spectral analysis.

The Fourier Transform is widely used in digital signal processing (DSP) due to its ability to decompose signals into sinusoidal components, which are easier to interpret and manipulate. This transformation is crucial in applications such as filtering, spectral analysis, and system identification. Furthermore, in statistical signal processing, the Fourier Transform plays a significant role in deriving power spectral densities, which describe how the power of a signal is distributed across different frequencies.

By transforming correlation and covariance functions into the frequency domain, we gain insights into the periodicity and spectral properties of signals, enabling efficient noise filtering, signal enhancement, and feature extraction. The following subsections explain the Fourier Transform of correlation and covariance functions, including cross-correlation and cross-covariance functions, along with their physical meanings.

Fourier Transform of Correlation and Covariance Functions

Fourier Transform of Autocorrelation Function

The autocorrelation function of a discrete-time signal $x[n]$ provides a measure of how the signal correlates with a time-shifted version of itself. In the frequency domain, this function reveals the power distribution of the signal over different frequency components. The Fourier Transform of the autocorrelation function is known as the Power Spectral Density (PSD) and is expressed as:

$$S_{XX}(e^{j\omega}) = \sum_{k=-\infty}^{\infty} R_{XX}[k]e^{-j\omega k} \tag{6.24}$$

where $S_{XX}(e^{j\omega})$ represents the power spectral density of the signal $x[n]$, and $R_{XX}[k]$ is the autocorrelation function. This relationship is established by the Wiener-Khinchin theorem, which states that the power spectral density is the Fourier Transform of the autocorrelation function.

The PSD provides a crucial tool for analyzing the power content of signals across different frequencies. This is particularly useful in applications such as noise reduction, speech processing, and radar signal analysis, where distinguishing between meaningful signals and noise components is essential.

Fourier Transform of Cross-Correlation Function

The cross-correlation function measures the similarity between two discrete-time signals $x[n]$ and $y[n]$ as a function of time lag. The Fourier Transform of the cross-correlation function is known as the cross-power spectral density and is given by:

$$S_{XY}(e^{j\omega}) = \sum_{k=-\infty}^{\infty} R_{XY}[k]e^{-j\omega k} \tag{6.25}$$

where $S_{XY}(e^{j\omega})$ describes how different frequency components of the two signals are correlated.

Cross-power spectral density is widely used in applications such as system identification and coherence analysis, where understanding how two signals interact in the frequency domain is critical. For example, in biomedical engineering, cross-correlation and its spectral analysis can help identify relationships between different physiological signals, such as heart rate and respiration signals.

Fourier Transform of Autocovariance Function

The autocovariance function is similar to the autocorrelation function but removes the effect of the mean of the signal.The Fourier Transform of the autocovariance function is given by:

$$S_{C,XX}(e^{j\omega}) = \sum_{k=-\infty}^{\infty} C_{XX}[k]e^{-j\omega k} \tag{6.26}$$

Since the mean of a signal does not contribute to its spectral content, the power spectral density derived from the autocovariance function provides a more accurate representation of fluctuations in signal power without mean interference. This is particularly useful in stochastic signal modeling, where second-order statistics are often more relevant than the raw signal itself.

Fourier Transform of Cross-Covariance Function

The Fourier Transform of the cross-covariance function of two signals $x[n]$ and $y[n]$, is:

$$S_{C,XY}(e^{j\omega}) = \sum_{k=-\infty}^{\infty} C_{XY}[k]e^{-j\omega k} \tag{6.27}$$

The cross-covariance function captures the statistical dependencies between two signals while eliminating mean effects. Its spectral representation is essential in multi-sensor signal processing, adaptive filtering, and communication systems, where understanding the correlation structure between signals is crucial for noise cancellation and data fusion.

Physical Meaning of the Transformed Functions

The Fourier Transform of correlation and covariance functions provides deep insights into signal properties can be listed as;

Power Spectral Density $S_{XX}(e^{j\omega})$: Describes how the power of a signal is distributed over frequency, highlighting dominant spectral components.
Cross-Power Spectral Density $S_{XY}(e^{j\omega})$: Indicates how two signals share frequency components, crucial for system identification and signal coherence analysis.
Spectral Density of Autocovariance $S_{C,XX}(e^{j\omega})$: Represents second-order statistics without mean interference, allowing for a cleaner spectral interpretation.
Spectral Density of Cross-Covariance $S_{C,XY}(e^{j\omega})$: Highlights the statistical dependencies between two signals in the frequency domain while removing mean effects.

These spectral functions form the basis of many engineering applications, such as radar signal processing, biomedical signal analysis, seismic signal processing, and speech enhancement. Their ability to reveal frequency-domain relationships makes them indispensable tools in both theoretical and applied signal processing.

6.6 Classification of Random Signals and Processes

Random signals and stochastic processes play a fundamental role in various engineering disciplines, particularly in signal processing, communications, and control systems. These signals and processes are characterized by inherent randomness, requiring statistical methods for analysis and interpretation. Understanding their classification helps in selecting appropriate tools for modeling, filtering, and prediction. The classification of random signals and processes can be made based on different criteria such as stationarity, ergodicity, independence, spectral properties, linearity, and Gaussianity.

Classification Based on Stationarity

Stationary Process

A random process $X[n]$ is called *stationary* if its statistical properties, such as mean and variance, remain constant over time. This implies that the behavior of the process does not change as time progresses, making it easier to analyze. A strictly stationary (SSS) process satisfies the condition:

$$P(X[n_1], X[n_2], \ldots, X[n_k]) = P(X[n_1 + m], X[n_2 + m], \ldots, X[n_k + m]) \tag{6.28}$$

for any time shift m and any set of time indices $\{n_1, n_2, \ldots, n_k\}$. However, strict stationarity is often too restrictive for practical applications.

Wide-Sense Stationary (WSS) Process

A process is *wide-sense stationary* (WSS) if it meets the following conditions:

The mean $E[X[n]]$ is constant and does not depend on n.

The autocorrelation function depends only on the time difference k, i.e.,

$$R_{XX}[m, n] = R_{XX}[m - n]. \tag{6.29}$$

WSS processes are widely used in signal processing due to their tractability and applicability in real-world scenarios.

Non-Stationary Process

If a process does not satisfy the stationarity conditions, it is called *non-stationary*. In such processes, the statistical properties change over time, making their analysis more complex. Non-stationary processes are common in speech signals, biomedical signals, and economic time series.

Classification Based on Ergodicity

A stochastic process is said to be *ergodic* if its statistical properties can be estimated from a single realization over a sufficiently long time interval. Ergodicity ensures that time averages and ensemble averages are equivalent, which is particularly useful in practical applications where only one realization of a process is available for analysis.

Mean Ergodic Process

A process is *mean ergodic* if the time average of the process converges to its expected mean value:

$$\lim_{T\to\infty} \frac{1}{T} \sum_{n=0}^{T} X[n] = E[X[n]]. \tag{6.30}$$

This means that if we observe the process over a sufficiently long duration, we can approximate its mean using a single realization.

Autocorrelation Ergodic Process

A process is *autocorrelation ergodic* if the time average of its autocorrelation function converges to its ensemble autocorrelation function:

$$\lim_{T\to\infty} \frac{1}{T} \sum_{n=0}^{T} X[n]X[n-k] = R_{XX}[k]. \tag{6.31}$$

Autocorrelation ergodicity ensures that statistical dependence between different time instants can be determined from a single realization, which is critical in spectral estimation and filtering applications.

Classification Based on Independence and Memory

Independent Process

A process is called *independent* if values at different time instances are statistically independent:

$$P(X[n_1], X[n_2]) = P(X[n_1])P(X[n_2]) \quad \text{for } n_1 \neq n_2. \tag{6.32}$$

Independent processes are important in modeling noise and interference in communication systems.

Markov Process

A Markov process satisfies the Markov property, which states that the future state depends only on the present state and not on past states:

$$P(X[n+1]|X[n], X[n-1], \ldots, X[0]) = P(X[n+1]|X[n]). \tag{6.33}$$

Markov processes are widely used in probabilistic modeling, speech recognition, and financial time series analysis.

Memoryless Process

A discrete-time memoryless process, often referred to as a memoryless sequence, is a stochastic process where the current state is independent of all past states, given the immediately preceding state. In simpler terms, the future behavior of the process depends only on its present state, and not on how it arrived at that state.

Let $X[n]$ be a discrete-time stochastic process, where n represents the discrete time index. The process $X[n]$ is memoryless if, for any $n \geq 1$, the conditional probability of $X[n]$ given the past states $X[n-1]$, $X[n-2]$, ..., $X[0]$ is equal to the conditional probability of $X[n]$ given only the immediate previous state $X[n-1]$:

$$\begin{aligned} &P(X[n] = x_n|X[n-1] = x_{n-1}, X[n-2] = x_{n-2}, ..., X[0] = x_0) \\ &= P(X[n] = x_n|X[n-1] = x_{n-1}) \end{aligned}$$

This equation implies that knowing the history of the process before $X[n-1]$ provides no additional information about the future state $X[n]$ beyond what is already contained in $X[n-1]$. In the realm of discrete-time stochastic processes, a special category exists known as memoryless processes, or memoryless sequences. These processes possess a unique characteristic: their future state is utterly independent of their past, provided their present state is known. Essentially, the process carries no "memory" of its history; only the immediate present matters. Mathematically, this means that the probability of the process transitioning to a particular state at a given time depends solely on its state at the immediately preceding time, and not on any earlier states.

A prime example of a memoryless process in discrete time is found in the geometric distribution. Imagine a series of coin flips, each independent of the last. The geometric distribution describes the number of flips required to obtain the first heads. The probability of getting heads on the next flip is unaffected by how many tails have already occurred, demonstrating the memoryless nature. This distribution stands as the discrete-time equivalent of the continuous exponential distribution, which also exhibits this lack of memory.

Moreover, memoryless processes form a specific subset of discrete-time Markov chains. While Markov chains dictate that the next state depends only on the current state, memoryless processes go further, ensuring complete independence from all past states. This stronger condition simplifies analysis, as it eliminates the need to track the entire history of the process. For instance, a simple random walk, where the direction of each step depends only on the current position, can be modeled as a memoryless process.

Furthermore, sequences of independent and identically distributed (i.i.d.) random variables are inherently memoryless. Each variable in such a sequence is independent of all others and shares the same probability distribution. A classic example is the repeated rolling of a fair die, where each roll is independent of the previous ones.

In essence, the memoryless property simplifies the analysis of discrete-time systems. However, it's crucial to acknowledge that many real-world systems do possess memory, meaning memoryless models may not always be suitable. Nevertheless, they provide a valuable tool for understanding and modeling systems where the past has no bearing on the future, given the present.

Classification Based on Power Spectral Density

White Noise Process

A white noise process has a constant power spectral density (PSD) over all frequencies:

$$S_{XX}(e^{j\omega}) = \sigma^2, \quad -\pi \leq \omega \leq \pi. \tag{6.34}$$

It exhibits no correlation between different time instances, making it a fundamental model for unpredictable signals.

Colored Noise Process

Pink Noise (1/*f* Noise)

Pink noise, also referred to as $1/f$ noise, is a type of noise where the power spectral density (PSD) is inversely proportional to the frequency. Mathematically, it is represented as:

$$S_{XX}(f) \propto \frac{1}{f^{\alpha}}, \quad 0 < \alpha < 2 \tag{6.35}$$

where typically $\alpha = 1$. This means that lower frequencies have more power than higher frequencies, giving pink noise a balanced spectral distribution across octaves.

Pink noise is commonly observed in biological systems, natural processes, and music. It is prevalent in heart rate variability, brain waves, financial markets, and climate patterns. Unlike white noise, pink noise has a more natural energy distribution, making it perceptually smoother to the human ear.

Pink noise is widely applied in many engineering and scientific disciplines due to its unique spectral properties. In audio engineering, it plays a crucial role in equalization, helping to balance frequencies in acoustic environments by mimicking the natural frequency response of human hearing. In scientific simulations, pink noise is often used to model stochastic behaviors in various fields such as physics, meteorology, and finance, where systems exhibit self-similar, fractal-like structures. Additionally, in speech and hearing research, pink noise is employed for auditory experiments, as its spectral characteristics closely resemble the way human perception processes sound, making it a valuable tool for analyzing speech intelligibility and auditory responses.

Brown Noise (Brownian Motion or Random Walk Noise)

Brown noise, also called Brownian noise, is a type of noise where the power spectral density decreases more steeply than in pink noise, typically as $1/f^2$. It is mathematically expressed as:

$$S_{XX}(f) \propto \frac{1}{f^2} \tag{6.36}$$

Brown noise derives its name from Brownian motion, which describes the random movement of particles in a fluid due to thermal energy. This connection highlights the fundamental nature of brown noise as a stochastic process with a cumulative effect over time. Unlike white noise, which has equal power across all frequencies, brown noise is generated by integrating white noise, resulting in a smoother waveform that emphasizes lower frequencies. This integration process causes the power spectral density of brown noise to decrease at a rate of $1/f^2$ leading to a much deeper, rumbling sound compared to pink noise.

Due to its strong low-frequency components and minimal high-frequency content, brown noise has several important applications. In oceanography and seismic analysis, it is used to model wave dynamics and seismic activity, aiding in the study of natural phenomena such as ocean currents and tectonic movements. Similarly, in fluid dynamics and turbulence studies, brown noise helps simulate real-world atmospheric and hydrodynamic behaviors, allowing for better understanding and prediction of complex fluid flows.

An illustrative comparison of these noises is shown in Fig. 6.1.

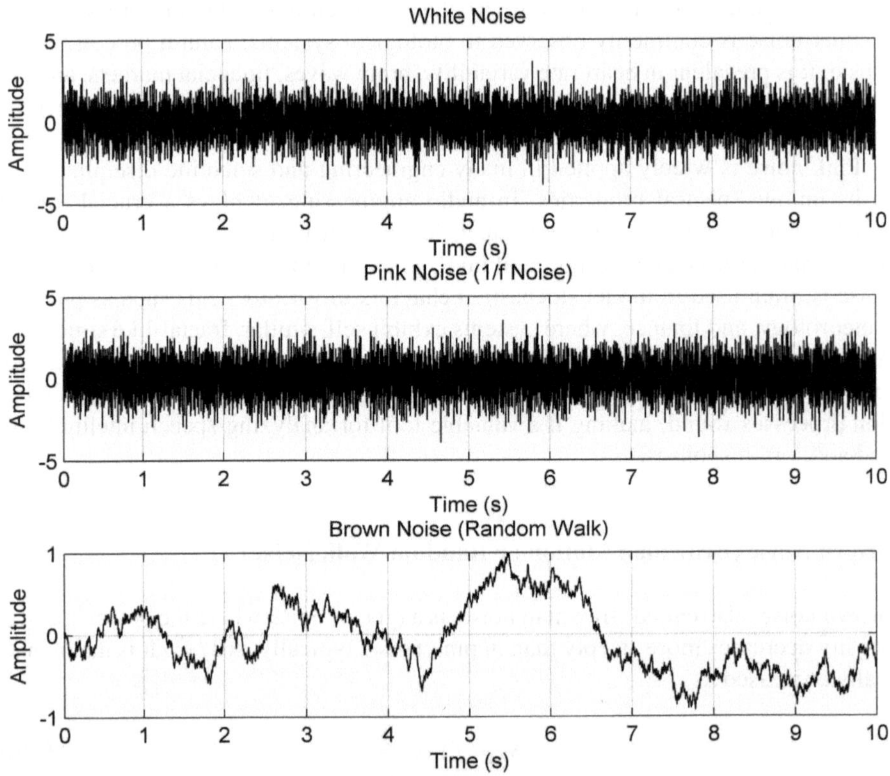

Fig. 6.1 Comparison of White, Pink, and Brown Noise

Classification of Random Signals and Processes

Random signals and processes are classified based on various criteria such as linearity, deterministic versus stochastic nature, and Gaussianity. These classifications help in analyzing and designing engineering systems where signal behavior impacts processing techniques such as filtering, estimation, and prediction.

Classification Based on Linearity

A fundamental way to classify random processes is by their linearity. The distinction between linear and nonlinear processes determines how they can be modeled and analyzed using mathematical tools such as convolution, Fourier transforms, and statistical techniques.

Linear Process

A process is called *linear* if its current value can be expressed as a linear combination of its past values and some noise component. Mathematically, a linear process is given by:

$$X[n] = \sum_{k=1}^{p} a_k X[n-k] + W[n], \tag{6.37}$$

where $X[n]$ represents the signal at time step n, the coefficients a_k determine the influence of past values, p denotes the order of the process, and $W[n]$ is a white noise sequence with zero mean and variance σ_W^2.

Two important classes of linear processes are the autoregressive (AR) process and the moving average (MA) process. In an autoregressive process, the current value is expressed as a function of past values with an added noise term:

$$X[n] = \sum_{k=1}^{p} a_k X[n-k] + W[n]. \tag{6.38}$$

On the other hand, a moving average process depends on past noise terms:

$$X[n] = \sum_{k=0}^{q} b_k W[n-k]. \tag{6.39}$$

Nonlinear Process

A process is *nonlinear* if its dependencies cannot be expressed using a linear combination of past values. Instead, the evolution of the process may involve nonlinear transformations such as multiplication, exponentiation, or other complex functional relationships:

$$X[n] = f(X[n-1], X[n-2], \dots) + W[n]. \tag{6.40}$$

Nonlinear models appear in various real-world applications, such as climate dynamics, biological systems, and financial markets.

Classification Based on Deterministic Versus Stochastic Nature

Another fundamental classification is whether a process is deterministic or stochastic, which determines predictability.

Deterministic Process

A process is *deterministic* if its future values can be precisely predicted from past values. A common example of a deterministic process is a sinusoidal signal given by:

$$X[n] = A\cos(\omega n + \phi), \tag{6.41}$$

where A represents the amplitude, ω is the angular frequency, and ϕ is the phase shift. Such signals frequently appear in communication and control systems where periodic behavior is observed.

Stochastic Process

A process is *stochastic* if its values follow a probability distribution and exhibit inherent randomness. The probability distribution function (PDF) of a stochastic process $X[n]$ is expressed as:

$$P(X[n] \leq x) = \int_{-\infty}^{x} f_X(t)dt, \tag{6.42}$$

where $f_X(t)$ is the probability density function. Stochastic processes commonly appear in diverse fields, such as Brownian motion in physics, white noise in signal processing, and Poisson processes in queuing theory.

Classification Based on Gaussianity

Gaussianity refers to whether a process follows a normal (Gaussian) distribution.

Gaussian Process

A process is *Gaussian* if all its finite-dimensional distributions are Gaussian. Mathematically, a Gaussian process satisfies:

$$X[n] \sim \mathcal{N}(\mu, \sigma^2), \tag{6.43}$$

where μ represents the mean and σ^2 denotes the variance. Gaussian processes play a fundamental role in statistical modeling, including applications such as Kalman filtering and Gaussian mixture models.

Non-Gaussian Process

A process is *non-Gaussian* if it does not follow a normal distribution. These processes often exhibit heavy-tailed or skewed distributions. Examples of non-Gaussian processes include impulsive noise environments modeled by alpha-stable distributions, speech signals, and certain economic time series. Since conventional Gaussian-based statistical methods may fail to capture the properties of such processes, advanced techniques such as higher-order statistics and wavelet analysis are often required.

The classification of random signals and processes into linear versus nonlinear, deterministic versus stochastic, and Gaussian versus non-Gaussian provides a structured approach to understanding their behavior. This classification is crucial for selecting appropriate analytical tools in engineering and scientific applications.

Problems

6.1 Let $X[n]$ be a wide-sense stationary (WSS) discrete-time random process with

$$E[X[n]] = 3,$$

$$R_{XX}[k] = 20 + 10e^{-|k|}$$

A new discrete-time random process is defined as:

$$Y[n] = \sum_{m=-\infty}^{\infty} X[m]$$

(a) Determine $E[Y[n]]$.
(b) Compute $\text{Var}[Y[n]]$.

6.2 Consider a discrete-time WSS random process $X[n]$ with

$$E[X[n]] = 5,$$

$$R_{XX}[k] = 30 + 12e^{-0.5|k|}$$

A new random process is generated as:

$$Y[n] = \sum_{m=-\infty}^{\infty} 2X[m-n]$$

(a) Compute the expected value $E[Y[n]]$.
(b) Find the variance $\text{Var}[Y[n]]$.

6.3 Consider a discrete-time Gaussian random process $X[n]$ with zero mean and the following autocorrelation function:

$$R_{XX}[k] = \frac{\sin(\pi k)}{\pi k}$$

Construct the covariance matrix for $k = 0, 1, \ldots, 8$.

6.4 Let $Y[n]$ be a discrete-time random process defined as:

$$Y[n] = A$$

where A is a discrete random variable having a uniform probability distribution over the set $\{-2, -1, 0, 1, 2\}$;

$$f_Y(y) = \begin{cases} \frac{1}{5} & y \in -2, -1, 0, 1, 2 \\ 0 & \text{otherwise} \end{cases}$$

(a) Calculate $\mu_Y[n]$.
(b) Derive the autocorrelation function, $R_{YY}[n_1, n_2]$.
(c) Is $Y[n]$ a wide-sense stationary (WSS) process?

6.5 Considering two statistically independent discrete-time random processes, $X[n]$ and $Y[n]$, then define a new random process $Z[n] = X[n]Y[n]$;

(a) Derive $\mu_Z[n]$ in terms of $\mu_X[n]$ and $\mu_Y[n]$.
(b) Find $R_{ZZ}[n_1, n_2]$ in terms of $R_{XX}[n_1, n_2]$ and $R_{YY}[n_1, n_2]$.

6.6 Consider a WSS discrete-time random process $X[n]$ with

$$E[X[n]] = 3$$
$$R_{XX}[k] = (0.5)^{|k|}$$

If $Y[n] = 4X[n]$;

(a) Find $\mu_Y[n]$.
(b) Define $R_{XY}[k]$.
(c) Define $R_{YY}[k]$.
(d) Is $Y[n]$ a WSS process?

6.7 Consider a WSS discrete-time random process $A[n]$ with

$$E[A[n]] = -1$$
$$R_{AA}[k] = 2e^{-|k|}$$

If $B[n] = -2A[n]$;

(a) Find $\mu_B[n]$.
(b) Define $R_{AB}[k]$.
(c) Define $R_{BB}[k]$.
(d) Is $B[n]$ a WSS process?

6.8 Let $X[n]$ be a WSS discrete-time random process with

$$R_{XX}[k] = e^{-|k|}$$

and $Y[n] = X[n+1] - X[n]$.

(a) Show that $\mu_Y[n] = 0$.
(b) Derive $R_{XY}[k]$.
(c) Derive $R_{YY}[k]$.
(d) Is $Y[n]$ a WSS process?

6.9 Let $X[n]$ be a WSS discrete-time random process with

$$R_{XX}[k] = (0.5)^{|k|}$$
$$\mu_X[n] = 2$$

A new random process is defined as

$$Y[n] = X[n] \cos(\omega n + \Phi)$$

where Φ is a discrete random variable uniformly distributed over the set $\{-\pi, -\pi + \frac{2\pi}{N}, \ldots, \pi - \frac{2\pi}{N}\}$, where N is a large integer.

(a) Find $\mu_Y[n]$.
(b) Derive $R_{YY}[k]$.
(c) Is $Y[n]$ a WSS process?

6.10 Let $X[n]$ be a WSS discrete-time random process with

$$R_{XX}[k] = 3(0.8)^{|k|}$$
$$\mu_X[n] = -1$$

A new random process is defined as

$$Z[n] = X[n]\sin(\omega n + \Psi)$$

where Ψ is a discrete random variable uniformly distributed over the set $\{-\pi, -\pi + \frac{2\pi}{M}, ..., \pi - \frac{2\pi}{M}\}$, where M is a large integer.

(a) Find $\mu_Z[n]$.
(b) Derive $R_{ZZ}[k]$.
(c) Is $Z[n]$ a WSS process?

6.11 Let $X[n]$ be a WSS discrete-time random process with mean μ_X and autocorrelation function $R_{XX}[k]$. A new random process $Y[n]$ is defined as:

$$Y[n] = X[n] + W[n]$$

where $W[n]$ is a WSS discrete-time random process, independent of $X[n]$, with mean μ_W and autocorrelation function $R_{WW}[k]$.

(a) Find the mean of $Y[n]$, $E[Y[n]]$.
(b) Find the autocorrelation function of $Y[n]$, $R_{YY}[k]$.
(c) Find the cross-correlation function between $X[n]$ and $Y[n]$, $R_{XY}[k]$.

6.12 Consider a WSS random process $X[n]$ with autocorrelation function $R_{XX}[k] = A\alpha^{|k|}$, where $A > 0$ and $0 < \alpha < 1$.

(a) Calculate the variance of $X[n]$.
(b) Determine the autocorrelation coefficient $\rho_{XX}[k] = \frac{R_{XX}[k]}{R_{XX}[0]}$.

6.13 A random process is defined as $X[n] = A\cos(\omega_0 n + \Theta)$, where A and ω_0 are constants, and Θ is a random variable uniformly distributed over $[0, 2\pi]$.

(a) Show that $X[n]$ is WSS.
(b) Find the autocorrelation function $R_{XX}[k]$.

6.14 Let $X[n]$ be a zero-mean WSS random process with autocorrelation function $R_{XX}[k]$. Consider the LTI system with impulse response $h[n] = a^n u[n]$, where $|a| < 1$. The output of the system when the input is $X[n]$ is $Y[n]$.

(a) Find the autocorrelation function $R_{YY}[k]$ of the output process $Y[n]$.
(b) Is $Y[n]$ WSS?

6.15 Two WSS random processes, $X[n]$ and $Y[n]$, are jointly WSS with autocorrelation functions $R_{XX}[k]$ and $R_{YY}[k]$, and cross-correlation function $R_{XY}[k]$. A new process $Z[n]$ is defined as $Z[n] = X[n] - Y[n]$.

(a) Find the autocorrelation function $R_{ZZ}[k]$ of $Z[n]$ in terms of $R_{XX}[k]$, $R_{YY}[k]$, and $R_{XY}[k]$.
(b) Find the cross-correlation function $R_{XZ}[k]$ between $X[n]$ and $Z[n]$.

6.16 A WSS random process $X[n]$ with mean μ_X is passed through a system with impulse response $h[n] = \{1, 2, 1\}$. The output is $Y[n]$.

(a) Find the mean of $Y[n]$, $E[Y[n]]$.
(b) Express the autocorrelation function $R_{YY}[k]$ of $Y[n]$ in terms of the autocorrelation function $R_{XX}[k]$ of $X[n]$.

6.17 Let $X[n]$ be a WSS random process with $E[X[n]] = 0$ and $R_{XX}[k] = \sigma^2\delta[k]$. Let $Y[n] = X[n] - X[n-1]$.

(a) Find the autocorrelation function $R_{YY}[k]$.
(b) Find the cross-correlation function $R_{XY}[k]$.

6.18 Consider a WSS process $X[n]$ with $R_{XX}[k] = (0.9)^{|k|}$. A new process $Y[n]$ is formed by $Y[n] = aX[n] + b$, where a and b are constants.

(a) Find the autocorrelation function of $Y[n]$, $R_{YY}[k]$.
(b) For what values of a and b is $Y[n]$ a zero-mean process?

6.19 A discrete-time random process is given by $X[n] = A + Bn$, where A and B are uncorrelated random variables with means μ_A and μ_B and variances σ_A^2 and σ_B^2.

(a) Find the mean function $E[X[n]]$.
(b) Find the autocorrelation function $R_{XX}[n_1, n_2]$.
(c) Is $X[n]$ WSS?

6.20 Let $X[n]$ be a WSS random process with autocorrelation function $R_{XX}[k]$. If $Y[n] = X[-n]$,

(a) Find the autocorrelation function $R_{YY}[k]$ of $Y[n]$.
(b) Find the cross-correlation function $R_{XY}[n_1, n_2]$ between $X[n]$ and $Y[n]$.
(c) Is $Y[n]$ WSS?

Reference

1. Monson H Hayes, *Statistical digital signal processing and modeling*, John Wiley & Sons, 1996.

Chapter 7
Discrete-Time Gaussian Random Processes

A Gaussian Random Process (GRP) is a special class of discrete-time stochastic processes in which every finite collection of random variables is jointly Gaussian distributed. This implies that for any finite set of time indices, the corresponding set of random variables possesses a multivariate normal distribution. Such a process provides a powerful and tractable framework for modeling uncertainty and randomness in a wide array of engineering, scientific, and mathematical applications.

A discrete-time Gaussian random process is formally defined as a collection of random variables indexed by the discrete-time variable $n \in \mathbb{Z}$, given by:

$$X[n] = \{X[n] : n \in \mathbb{Z}\} \tag{7.1}$$

This definition implies that for any finite sequence of time indices $n_1, n_2, \ldots, n_N$, the corresponding random vector $(X[n_1], X[n_2], \ldots, X[n_N])$ adheres to a multivariate Gaussian distribution. This key property distinguishes GRPs from other types of stochastic processes and significantly simplifies their theoretical treatment and practical implementation.

Gaussian processes are entirely characterized by their first and second-order statistical moments. That is, complete knowledge of the mean and autocovariance functions of the process suffices to describe its probabilistic behavior. These two functions are defined as follows:

The mean function, which provides the expected value of the process at each time index:

$$\mu_X[n] = \mathbb{E}[X[n]] \tag{7.2}$$

This function captures the deterministic trend or central tendency around which the random process fluctuates.

The autocovariance function, which quantifies the linear dependence and temporal correlation between the values of the process at different time indices:

$$R_X[n_1, n_2] = \mathbb{E}[(X[n_1] - \mu_X[n_1])(X[n_2] - \mu_X[n_2])] \tag{7.3}$$

M. Catak and T. Allahviranloo, *Discrete Random Processes*, Studies in Systems, Decision and Control 644, https://doi.org/10.1007/978-3-032-08630-3_7

The autocovariance function not only encodes the correlation structure of the process but also plays a central role in determining its spectral and filtering properties.

Due to the analytical tractability of the Gaussian distribution, GRPs are of paramount importance in stochastic modeling and signal processing. The assumption of Gaussianity leads to closed-form expressions in many fundamental problems, including optimal estimation (e.g., Wiener and Kalman filtering), hypothesis testing, detection theory, and linear prediction. Moreover, the Gaussian nature allows for the application of powerful tools from linear algebra and functional analysis.

In engineering applications, discrete-time Gaussian random processes are routinely employed to model thermal and electronic noise in circuits, unpredictable fluctuations in communication channels, and measurement uncertainties in control systems. They provide the statistical foundation for the design of robust signal processing algorithms and adaptive filtering techniques. In addition, the stationarity or wide-sense stationarity (WSS) assumptions imposed on GRPs lead to further simplifications, especially in spectral analysis, where the power spectral density becomes a critical descriptor of the frequency content of the process.

Furthermore, Gaussian processes exhibit properties such as closure under linear operations and conditioning, which make them particularly useful in Bayesian frameworks and machine learning. The ability to condition a Gaussian process on observations yields another Gaussian process, facilitating elegant formulations of posterior distributions and predictive models.

7.1 Mathematical Properties of Discrete-Time Gaussian Random Processes

Gaussian random processes have a wide range of mathematical properties because of their analytical tractability as well as clearly defined statistical structure. These properties facilitate a deeper understanding of how such processes behave under transformations, how their dependencies are structured, and how they can be manipulated in signal processing and estimation contexts. In this section, we elaborate on the fundamental concepts of stationarity, correlation structures, and cross-statistics.

7.1.1 Stationarity

Stationarity is a crucial concept in the analysis of stochastic processes, describing whether and how the statistical properties of a process vary with time. For discrete-time Gaussian random processes, stationarity comes in two main forms:

Strict-Sense Stationarity (SSS): A process $\{X[n]\}$ is said to be strictly stationary if the joint distribution of any collection of samples remains invariant under time shifts. That is, for all integers $n_1, n_2, \ldots, n_k$ and for all $m \in \mathbb{Z}$,

$$(X[n_1], X[n_2], \ldots, X[n_k]) \stackrel{d}{=} (X[n_1 + m], X[n_2 + m], \ldots, X[n_k + m])$$

where $\stackrel{d}{=}$ denotes equality in distribution. This implies that the probabilistic behavior of the process is independent of absolute time, depending only on relative time differences.

Wide-Sense Stationarity (WSS): A weaker, yet commonly used form of stationarity, particularly in engineering applications. A process $\{X[n]\}$ is WSS if it satisfies the following two conditions:

The mean function is constant:

$$\mu_X[n] = \mathbb{E}[X[n]] = \mu, \quad \forall n$$

The autocovariance function depends only on the time difference (or lag) $k = n_2 - n_1$, not on the absolute times:

$$R_X[n_1, n_2] = R_X[k] \tag{7.4}$$

For Gaussian processes, WSS implies that second-order statistics are shift-invariant, which greatly simplifies their analysis and modeling, especially in spectral and filtering domains.

7.1.2 *Correlation and Covariance Matrices*

In practice, one often considers a finite collection of samples $X[n_1], X[n_2], \ldots, X[n_N]$ from a discrete-time Gaussian random process. The joint statistical structure of these samples is described using correlation and covariance matrices. These matrices are fundamental in multivariate analysis, providing compact representations of the relationships between random variables.

Autocorrelation Matrix

The autocorrelation matrix $\boldsymbol{R}_X \in \mathbb{C}^{N \times N}$ is defined as:

$$\boldsymbol{R}_X = \begin{bmatrix} R_X[0] & R_X[1] & R_X[2] & \dots & R_X[N-1] \\ R_X[-1] & R_X[0] & R_X[1] & \dots & R_X[N-2] \\ R_X[-2] & R_X[-1] & R_X[0] & \dots & R_X[N-3] \\ \vdots & \vdots & \vdots & \ddots & \vdots \\ R_X[-(N-1)] & R_X[-(N-2)] & R_X[-(N-3)] & \dots & R_X[0] \end{bmatrix} \tag{7.5}$$

Each entry $R_X[k]$ in this matrix corresponds to the autocorrelation at lag k. For WSS Gaussian processes, this matrix is Toeplitz and Hermitian (i.e., $\boldsymbol{R}_X = \boldsymbol{R}_X^H$), which is advantageous in signal processing algorithms such as Wiener filtering.

Autocovariance Matrix

The autocovariance matrix $\boldsymbol{C}_X$ adjusts for the mean of the process and is given by:

$$\boldsymbol{C}_X = \boldsymbol{R}_X - \mu_X \mu_X^T \tag{7.6}$$

where $\mu_X \in \mathbb{R}^N$ is the mean vector:

$$\mu_X = \left[\mu_X[n_1] \; \mu_X[n_2] \; \dots \; \mu_X[n_N]\right]^T$$

The autocovariance matrix captures the second-order dependencies between variables after removing the influence of the mean. It is symmetric (or Hermitian in the complex case) and positive semi-definite, which makes it a key component in eigenvalue analysis and principal component techniques.

7.1.3 *Cross-Correlation and Cross-Covariance*

When analyzing two jointly distributed discrete-time random processes $X[n]$ and $Y[n]$, it is often necessary to study their mutual dependence. This is achieved through the cross-correlation and cross-covariance functions and their matrix representations.

Cross-Correlation Function

The cross-correlation function between $X[n]$ and $Y[n]$ is defined as:

$$R_{XY}[n_1, n_2] = \mathbb{E}[X[n_1]Y^*[n_2]] \tag{7.7}$$

This function measures the similarity between $X[n]$ and a time-shifted version of $Y[n]$. When the processes are WSS, $R_{XY}[n_1, n_2]$ depends only on the lag $k = n_2 - n_1$, and we denote it as $R_{XY}[k]$.

Cross-Covariance Function

The cross-covariance function removes the mean effects and is given by:

$$C_{XY}[n_1, n_2] = R_{XY}[n_1, n_2] - \mu_X[n_1]\mu_Y^*[n_2] \tag{7.8}$$

This function captures the linear relationship between the fluctuations of $X[n]$ and $Y[n]$ about their respective means. Like the autocovariance function, it is essential in the study of estimation and system identification, where one process is used to infer or predict the behavior of another.

The matrix forms of these functions are particularly useful when dealing with multivariate Gaussian vectors, and they serve as the basis for linear prediction theory, optimal filtering, and signal decorrelation techniques.

Example 7.1

Consider the random samples given in Table 7.1 where each column is a random sequence. The random samples generated from a Gaussian distribution with $\mu = 25$, and $\sigma = 5$ using GNU Octave.

(a) Calculate the sample mean, $\overline{\mu_X}$
(b) Calculate the sample standard deviation, σ_X
(c) Find the coefficient of variation, c_v, for each random sequence.
(d) Derive the autocorrelation matrix, R_{XX}

Table 7.1 Random samples generated from Gaussian distribution with $\mu = 25$, and $\sigma = 5$

X1	X2	X3	X4	X5
26.5	22.1	31.2	28.7	21.8
30.1	28.5	24.3	23.9	27.6
23.8	25.9	27.8	30.5	22.1
21.2	29.7	22.5	26.1	31.4
28.9	20.3	29.1	24.8	25.7
24.6	27.1	23.4	32.0	29.3
27.3	23.5	26.7	21.5	20.9
22.7	31.8	20.9	29.4	26.5
31.5	24.2	28.3	25.2	23.1
25.4	26.8	21.7	31.1	28.8

Solution 7.1

(a) Let $n = 10$ be the number of samples in each sequence.

$$\overline{\mu_{X1}} = \frac{26.5+30.1+23.8+21.2+28.9+24.6+27.3+22.7+31.5+25.4}{10} = \frac{262}{10} = 26.2$$

$$\overline{\mu_{X2}} = \frac{22.1+28.5+25.9+29.7+20.3+27.1+23.5+31.8+24.2+26.8}{10} = \frac{260}{10} = 26.0$$

$$\overline{\mu_{X3}} = \frac{31.2+24.3+27.8+22.5+29.1+23.4+26.7+20.9+28.3+21.7}{10} = \frac{255.9}{10} = 25.59$$

$$\overline{\mu_{X4}} = \frac{28.7+23.9+30.5+26.1+24.8+32.0+21.5+29.4+25.2+31.1}{10} = \frac{273}{10} = 27.3$$

$$\overline{\mu_{X5}} = \frac{21.8+27.6+22.1+31.4+25.7+29.3+20.9+26.5+23.1+28.8}{10} = \frac{257.2}{10} = 25.72$$

(b) The sample standard deviation can be calculated for each column using the formula:

$$\sigma_X = \sqrt{\frac{\sum_{i=1}^{n}(x_i - \overline{\mu_X})^2}{n-1}}$$

For X_1:
$\sum_{i=1}^{10}(x_{1i} - 26.2)^2 = 98.1$

$$\sigma_{X1} = \sqrt{\frac{98.1}{10-1}} = \sqrt{\frac{98.1}{9}} \approx \sqrt{10.9} \approx 3.30$$

For X_2:
$\sum_{i=1}^{10}(x_{2i} - 26.0)^2 = 112.43$

$$\sigma_{X2} = \sqrt{\frac{112.43}{9}} \approx \sqrt{12.49} \approx 3.53$$

For X_3:
$\sum_{i=1}^{10}(x_{3i} - 25.59)^2 = 110.389$

$$\sigma_{X3} = \sqrt{\frac{110.389}{9}} \approx \sqrt{12.27} \approx 3.50$$

For X_4:
$\sum_{i=1}^{10}(x_{4i} - 27.3)^2 = 110.44$

$$\sigma_{X4} = \sqrt{\frac{110.44}{9}} \approx \sqrt{12.27} \approx 3.50$$

For X_5:
$\sum_{i=1}^{10}(x_{5i} - 25.72)^2 = 117.276$

$$\sigma_{X5} = \sqrt{\frac{117.276}{9}} \approx \sqrt{13.03} \approx 3.61$$

(c) The coefficient of variation is given by:

$$c_v = \frac{\sigma_X}{|\mu_X|}$$

$c_{v1} = \frac{3.30}{|26.2|} \approx 0.126$

$c_{v2} = \frac{3.53}{|26.0|} \approx 0.136$

$c_{v3} = \frac{3.50}{|25.59|} \approx 0.137$

$c_{v4} = \frac{3.50}{|27.3|} \approx 0.128$

$c_{v5} = \frac{3.61}{|25.72|} \approx 0.140$

(d) The autocorrelation matrix R_{xx} can be estimated as:

$$R_{XX} \approx C_{XX} + \overline{\mu_X \mu_X}^T$$

where the sample covariance matrix C_{xx} is:

$$C_{ij} = \frac{1}{n-1} \sum_{k=1}^{n} (x_{ik} - \overline{\mu_X}_i)(x_{jk} - \overline{\mu_X}_j)$$

Calculating the elements of C_{XX}: $C_{11} \approx 10.90$, $C_{22} \approx 12.49$, $C_{33} \approx 12.27$, $C_{44} \approx 12.27$, $C_{55} \approx 13.03$
$C_{12} = C_{21} \approx -6.66$
$C_{13} = C_{31} \approx 3.86$
$C_{14} = C_{41} \approx -2.24$
$C_{15} = C_{51} \approx -4.06$
$C_{23} = C_{32} \approx -3.33$
$C_{24} = C_{42} \approx 2.58$
$C_{25} = C_{52} \approx 4.33$
$C_{34} = C_{43} \approx -4.49$
$C_{35} = C_{53} \approx -2.86$
$C_{45} = C_{54} \approx 1.54$

So, the sample covariance matrix is approximately:

$$C \approx \begin{bmatrix} 10.90 & -6.66 & 3.86 & -2.24 & -4.06 \\ -6.66 & 12.49 & -3.33 & 2.58 & 4.33 \\ 3.86 & -3.33 & 12.27 & -4.49 & -2.86 \\ -2.24 & 2.58 & -4.49 & 12.27 & 1.54 \\ -4.06 & 4.33 & -2.86 & 1.54 & 13.03 \end{bmatrix}$$

The outer product of the means is:

$$\overline{\mu_X \mu_X}^T = \begin{bmatrix} 26.2 \\ 26.0 \\ 25.59 \\ 27.3 \\ 25.72 \end{bmatrix} \begin{bmatrix} 26.2 & 26.0 & 25.59 & 27.3 & 25.72 \end{bmatrix} = \begin{bmatrix} 686.44 & 681.20 & 670.46 & 714.26 & 673.86 \\ 681.20 & 676.00 & 665.34 & 709.80 & 669.72 \\ 670.46 & 665.34 & 654.84 & 698.67 & 658.81 \\ 714.26 & 709.80 & 698.67 & 745.29 & 702.28 \\ 673.86 & 669.72 & 658.81 & 702.28 & 661.52 \end{bmatrix}$$

The autocorrelation matrix R_{xx} is approximately:

$$R_{XX} \approx C_{XX} + \overline{\mu_X \mu_X}^T = \begin{bmatrix} 697.34 & 674.54 & 674.32 & 712.02 & 669.80 \\ 674.54 & 688.49 & 662.01 & 712.38 & 674.05 \\ 674.32 & 662.01 & 667.11 & 694.18 & 655.95 \\ 712.02 & 712.38 & 694.18 & 757.56 & 703.82 \\ 669.80 & 674.05 & 655.95 & 703.82 & 674.55 \end{bmatrix}$$

7.2 Fourier Transform of Correlation and Covariance Functions

The Fourier transform plays a central role in the spectral analysis of random processes, particularly in understanding how signal power and mutual dependencies are distributed across frequency. The correlation and covariance functions of stochastic processes in the time domain can be transformed into their spectral counterparts in the frequency domain using the Fourier transform. This transformation not only facilitates the analysis of random signals but also forms the foundation for many practical applications in signal processing, such as filtering, modulation, and system identification.

7.2.1 Power Spectral Density (PSD)

The Power Spectral Density (PSD) quantifies the distribution of power of a wide-sense stationary (WSS) random process across different frequency components. According to the **Wiener-Khinchin Theorem**, the PSD of a discrete-time random process $X[n]$ is defined as the discrete-time Fourier transform (DTFT) of its autocorrelation function $R_X[k]$:

$$S_X(e^{j\omega}) = \sum_{k=-\infty}^{\infty} R_X[k] e^{-j\omega k} \tag{7.9}$$

This spectral representation provides a powerful tool for analyzing and designing systems in the frequency domain. For real-valued processes, $S_X(e^{j\omega})$ is a real and even function of ω, and it is non-negative for all $\omega \in [-\pi, \pi]$. The total average power of the process can be obtained by integrating the PSD over the normalized frequency domain:

$$\text{Power} = \frac{1}{2\pi}\int_{-\pi}^{\pi} S_X(e^{j\omega})d\omega \tag{7.10}$$

7.2.2 *Cross Power Spectral Density (CPSD)*

The Cross Power Spectral Density (CPSD), also called the cross-spectral density, describes how two jointly WSS random processes $X[n]$ and $Y[n]$ are correlated in the frequency domain. It is the DTFT of the cross-correlation function $R_{XY}[k]$:

$$S_{XY}(e^{j\omega}) = \sum_{k=-\infty}^{\infty} R_{XY}[k]e^{-j\omega k} \tag{7.11}$$

The CPSD is generally complex-valued and provides insight into the phase and amplitude relationships between the frequency components of $X[n]$ and $Y[n]$. When $X[n] = Y[n]$, the CPSD reduces to the PSD.

Example 7.2

Consider a discrete-time WSS process $X[n]$ with the following autocorrelation function:

$$R_X[k] = \begin{cases} 4, & k = 0 \\ 2, & |k| = 1 \\ 1, & |k| = 2 \\ 0, & \text{otherwise} \end{cases}$$

Compute the power spectral density $S_X(e^{j\omega})$.

Solution 7.2

We apply the DTFT of $R_X[k]$ as follows:

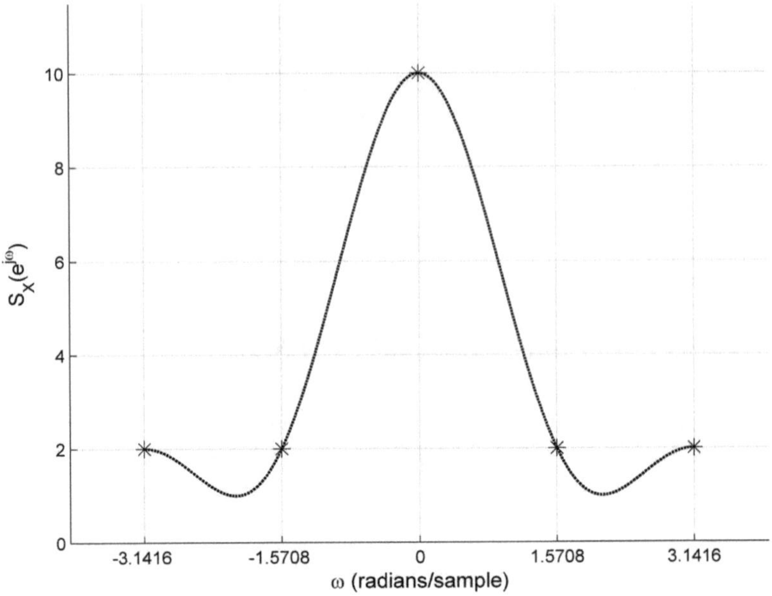

Fig. 7.1 Power spectral density of Example 7.2

$$
\begin{aligned}
S_X(e^{j\omega}) &= \sum_{k=-\infty}^{\infty} R_X[k]e^{-j\omega k} \\
&= R_X[0] + R_X[1]e^{-j\omega} + R_X[-1]e^{j\omega} + R_X[2]e^{-j2\omega} + R_X[-2]e^{j2\omega} \\
&= 4 + 2e^{-j\omega} + 2e^{j\omega} + 1e^{-j2\omega} + 1e^{j2\omega}
\end{aligned}
$$

Since the autocorrelation is real and symmetric, the PSD is real-valued:

$$
\begin{aligned}
S_X(e^{j\omega}) &= 4 + 2\left(e^{-j\omega} + e^{j\omega}\right) + \left(e^{-j2\omega} + e^{j2\omega}\right) \\
&= 4 + 4\cos(\omega) + 2\cos(2\omega)
\end{aligned}
$$

$$
S_X(e^{j\omega}) = 4 + 4\cos(\omega) + 2\cos(2\omega) \tag{7.12}
$$

This result provides the spectral content of the random process. The presence of cosine terms indicates that the process contains power at the corresponding frequencies. The PSD is plotted to visualize how the energy is distributed over $\omega \in [-\pi, \pi]$ as shown in Fig. 7.1.

7.3 Engineering Applications of Discrete-Time Gaussian Random Processes

7.3.1 Application 1: Gaussian Noise in Digital Communication

In a digital communication system, a transmitted signal is affected by noise, which is typically modeled as Additive White Gaussian Noise (AWGN). The received signal can be represented as:

$$Y[n] = X[n] + W[n]$$

where $X[n]$ is the transmitted signal; $W[n] \sim \mathcal{N}(0, \sigma^2)$ is the AWGN with zero mean and variance σ^2; $Y[n]$ is the received signal.

Signal-to-Noise Ratio (SNR) is a crucial metric in communication systems, defined as:

$$\text{SNR} = \frac{P_X}{\sigma^2}$$

where $P_X = E[|X[n]|^2]$ is the average power of the transmitted signal, and σ^2 is the noise variance.

Assume a Binary Phase Shift Keying (BPSK) modulation, where $X[n]$ takes values $\pm A$. Let $A = 1$, so the signal power is:

$$P_X = E[X[n]^2] = A^2 = 1$$

Suppose the noise variance is $\sigma^2 = 0.1$.

Now, the SNR in dB is;

$$\text{SNR}_{\text{dB}} = 10 \log_{10}\left(\frac{P_X}{\sigma^2}\right) = 10 \log_{10}\left(\frac{1}{0.1}\right)$$

$$= 10 \log_{10}(10) = 10 \text{ dB}$$

The probability of bit error for BPSK under AWGN is given by:

$$P_e = Q\left(\sqrt{\frac{2P_X}{\sigma^2}}\right)$$

where $Q(x)$ is the Q-function, defined as;

$$Q(x) = \frac{1}{\sqrt{2\pi}} \int_x^{\infty} e^{-t^2/2} dt$$

For SNR = 10 dB:

$$P_e = Q\left(\sqrt{\frac{2 \times 1}{0.1}}\right) = Q(\sqrt{20})$$

Using numerical approximations, $Q(\sqrt{20}) \approx 2 \times 10^{-5}$, meaning that only 2 bits out of 100,000 are expected to be received in error. If the noise variance σ^2 increases, the BER increases, meaning more errors occur. While SNR improves (e.g., by increasing transmission power P_X), the BER decreases, leading to more reliable communication.

7.3.2 Application 2: Kalman Filtering in Discrete-Time Systems

The Kalman filter is widely used in control and signal estimation problems.

A linear discrete-time system with Gaussian noise is modeled as:

$$X_{k+1} = AX_k + BU_k + W_k, \quad W_k \sim \mathcal{N}(0, Q) \tag{7.13}$$

$$Y_k = CX_k + V_k, \quad V_k \sim \mathcal{N}(0, R) \tag{7.14}$$

where:

X_k is the state vector at time step k.
U_k is the control input.
Y_k is the measurement.
W_k and V_k are the process and measurement noise, respectively, both assumed to be Gaussian with covariance matrices Q and R.

The Kalman filter recursively updates the estimated state $\hat{X}_k$ using:

$$\hat{X}_{k|k} = \hat{X}_{k|k-1} + K_k(Y_k - C\hat{X}_{k|k-1}) \tag{7.15}$$

where K_k is the Kalman gain:

$$K_k = P_{k|k-1}C^T(CP_{k|k-1}C^T + R)^{-1} \tag{7.16}$$

and $P_{k|k-1}$ is the predicted state covariance matrix.

Example 7.3: Kalman Filter for Position and Velocity Estimation

Consider a discrete-time tracking system where the state vector at time step k is:

$$X_k = \begin{bmatrix} p_k \\ v_k \end{bmatrix},$$

with p_k representing the position and v_k the velocity of an object.

The system follows the linear state-space model:

$$X_{k+1} = AX_k + BU_k + W_k,$$
$$Y_k = HX_k + V_k,$$

where:

$$A = \begin{bmatrix} 1 & \Delta t \\ 0 & 1 \end{bmatrix}, \quad B = \begin{bmatrix} 0 \\ 0 \end{bmatrix}, \quad H = \begin{bmatrix} 1 & 0 \end{bmatrix}.$$

Here, $\Delta t = 1$ s is the sampling interval, $W_k \sim \mathcal{N}(0, Q)$ is the process noise, and $V_k \sim \mathcal{N}(0, R)$ is the measurement noise, with:

$$Q = \begin{bmatrix} 0.1 & 0 \\ 0 & 0.01 \end{bmatrix}, \quad R = 1.$$

Initial Conditions

Assume the initial state and estimation parameters are:

$$\hat{X}_0 = \begin{bmatrix} 0 \\ 1 \end{bmatrix}, \quad P_0 = \begin{bmatrix} 1 & 0 \\ 0 & 1 \end{bmatrix}.$$

Kalman Filter Equations

The Kalman filter consists of the following two stages:

Prediction Step:

$$\hat{X}_{k|k-1} = A\hat{X}_{k-1|k-1},$$
$$P_{k|k-1} = AP_{k-1|k-1}A^T + Q.$$

Update Step:

$$K_k = P_{k|k-1} H^T (H P_{k|k-1} H^T + R)^{-1},$$
$$\hat{X}_{k|k} = \hat{X}_{k|k-1} + K_k (Y_k - H \hat{X}_{k|k-1}),$$
$$P_{k|k} = (I - K_k H) P_{k|k-1}.$$

Numerical Simulation for 3 Steps

Let us assume the true state evolves as follows (no input applied, but process noise exists):

$$X_0 = \begin{bmatrix} 0 \\ 1 \end{bmatrix}, \quad X_1 = \begin{bmatrix} 1.1 \\ 0.95 \end{bmatrix}, \quad X_2 = \begin{bmatrix} 2.0 \\ 0.90 \end{bmatrix}, \quad X_3 = \begin{bmatrix} 2.85 \\ 0.85 \end{bmatrix}.$$

Assume the noisy position measurements are:

$$Y_1 = 1.3, \quad Y_2 = 1.7, \quad Y_3 = 3.1.$$

We apply the Kalman filter recursively:
At $k = 1$:

$$\hat{X}_{1|0} = \begin{bmatrix} 1 & 1 \\ 0 & 1 \end{bmatrix} \begin{bmatrix} 0 \\ 1 \end{bmatrix} = \begin{bmatrix} 1 \\ 1 \end{bmatrix},$$
$$P_{1|0} = A P_0 A^T + Q = \begin{bmatrix} 2.1 & 1 \\ 1 & 1.01 \end{bmatrix},$$
$$K_1 = P_{1|0} H^T (H P_{1|0} H^T + R)^{-1} = \begin{bmatrix} 2.1 \\ 1 \end{bmatrix} \cdot \frac{1}{2.1 + 1} = \begin{bmatrix} 0.6774 \\ 0.3226 \end{bmatrix},$$
$$\hat{X}_{1|1} = \hat{X}_{1|0} + K_1 (Y_1 - H \hat{X}_{1|0}) = \begin{bmatrix} 1 \\ 1 \end{bmatrix} + \begin{bmatrix} 0.6774 \\ 0.3226 \end{bmatrix} (1.3 - 1) = \begin{bmatrix} 1.2032 \\ 1.0968 \end{bmatrix}.$$

At $k = 2$:
Repeat the steps using $\hat{X}_{1|1}$ and $P_{1|1}$ to get $\hat{X}_{2|2} \approx \begin{bmatrix} 1.8059 \\ 1.0101 \end{bmatrix}$.
At $k = 3$:
By means of proceeding similarly, then the filter produces:

$$\hat{X}_{3|3} \approx \begin{bmatrix} 2.765 \\ 0.924 \end{bmatrix}.$$

The Kalman filter effectively smooths out the noisy position observations while simultaneously estimating the unmeasured velocity component of the state vector. The position estimates produced by the filter tend to converge toward the true values; for instance, the estimated position at time step 3 is $\hat{p}_3 = 2.765$, which is close to the actual position $p_3 = 2.85$. Even though the velocity is not directly observed, the filter is able to infer its value based on the changes in position over time and the known system dynamics. Furthermore, as more measurements are incorporated, the filter reduces the uncertainty in its estimates, which is reflected by a general decrease in the state covariance matrix P_k.

This example demonstrates the power of the Kalman filter in estimating both directly observed and unobserved states in the presence of noise. It highlights how optimal estimation techniques are crucial in tracking applications such as radar, navigation, and robotics.

7.3.3 Application 3: Financial Modeling with Gaussian Processes

Gaussian processes are effectively employed in financial modeling, particularly in the prediction of stock prices, risk assessment, and pricing of financial derivatives. In discrete-time, we denote the stock price at day n as $S[n]$, and model the prices as a realization of a Gaussian process:

$$S[n] \sim \mathcal{GP}(m[n], k[n, n'])$$

where $m[n]$ is the mean function, and $k[n, n']$ is the covariance kernel defining the similarity between prices at days n and n'.

Suppose that the stock prices were observed over 10 consecutive days:

Day n	1	2	3	4	5	6	7	8	9	10
$S[n]$	100	102	105	103	106	108	110	109	112	115

We aim to predict $S[11]$ using the Gaussian process regression model with the squared exponential kernel:

$$k[n, n'] = \sigma_f^2 \exp\left(-\frac{(n-n')^2}{2\ell^2}\right)$$

Assume hyperparameters $\sigma_f^2 = 100$, $\ell = 1.5$, and observation noise variance $\sigma_n^2 = 1$.

Using GP regression, the predicted mean and variance of $S[11]$ are computed as:

$$\mu[11] = \boldsymbol{k}_*^\top (K + \sigma_n^2 I)^{-1} \boldsymbol{y} \tag{7.17}$$

$$\sigma^2[11] = k[11, 11] - \boldsymbol{k}_*^\top (K + \sigma_n^2 I)^{-1} \boldsymbol{k}_* \tag{7.18}$$

where $\boldsymbol{k}_*$ is the vector of covariances between $S[11]$ and the training days $1 \leq n \leq 10$, and K is the covariance matrix of training inputs.

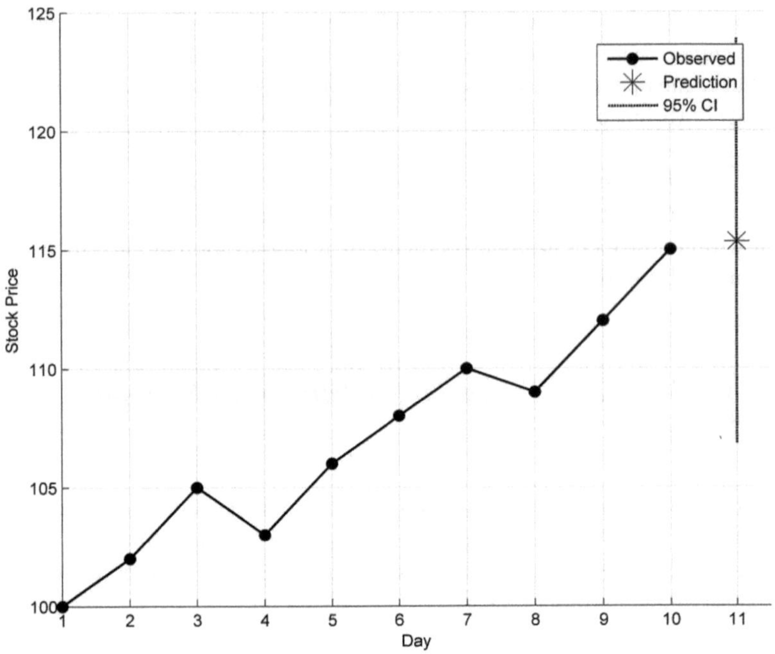

Fig. 7.2 An illustrative example for Gaussian process forecasting of stock prices

The result yields a predicted stock price on day 11 of approximately $\mu[11] \approx 117.6$, with a 95% confidence interval of approximately $[114.9,\ 120.3]$.

An illustrative example for the application of Gaussian random process in order to estimate stock prices is shown in Fig. 7.2.

This example illustrates how Gaussian processes can model both the trend and uncertainty in stock prices. The predicted value lies in line with the overall upward trend, and the confidence interval accounts for inherent market volatility. As more data points are incorporated, the confidence bounds can become narrower, enhancing prediction certainty.

Problems

7.1 A discrete-time Gaussian random process X[n] has a constant mean $\mu_X = 2$ and an autocovariance function $R_X[k] = 5 \cdot (0.7)^{|k|}$.

(a) Calculate the variance of the process X[n].
(b) Compute the autocorrelation matrix R_X for $k = -2, -1, 0, 1, 2$.

7.2 Two jointly Gaussian random processes, X[n] and Y[n], have zero means and the following autocorrelation and cross-correlation functions: $R_X[k] = 9 \cdot (0.4)^{|k|}$,

$R_Y[k] = 2\delta[k]$, $R_{XY}[k] = 3 \cdot (0.5)^k u[k]$, where $\delta[k]$ is the Kronecker delta function and $u[k]$ is the unit step function.

(a) Calculate the cross-correlation function $R_{YX}[k]$.
(b) Compute the cross-covariance function $C_{XY}[k]$.

7.3 A WSS Gaussian random process X[n] has the autocorrelation function $R_X[k] = 8\cos(0.2\pi k) + 3\delta[k]$.

(a) Calculate the power spectral density (PSD) $S_X(e^{j\omega})$.
(b) Determine the total average power of the process.

7.4 Given a discrete-time WSS Gaussian process X[n] with mean $\mu_X = -1$ and autocorrelation function $R_X[k] = 2 + (0.8)^{|k|}$, find the autocovariance matrix C_X for $n = 0, 1, 2$.

7.5 A digital communication system transmits a signal X[n] through an AWGN channel. The received signal is Y[n] = X[n] + W[n], where W[n] is AWGN with zero mean and variance $\sigma^2 = 0.5$. If the transmitted signal X[n] has power $P_X = 10$, calculate the SNR (Signal-to-Noise Ratio) in dB.

7.6 In a Kalman filter, the state vector is updated as $\hat{X}_{k|k} = \hat{X}_{k|k-1} + K_k(Y_k - H\hat{X}_{k|k-1})$, where K_k is the Kalman gain. Given $P_{k|k-1} = \begin{bmatrix} 3 & 1 \\ 1 & 2 \end{bmatrix}$, $H = \begin{bmatrix} 1 & 1 \end{bmatrix}$, and $R = 1$, calculate the Kalman gain K_k.

7.7 A discrete-time random process is defined as $Y[n] = 2X[n] + X[n-1]$, where X[n] is a WSS Gaussian process with mean μ_X and autocorrelation function $R_X[k]$. Find the autocorrelation function $R_Y[k]$ in terms of $R_X[k]$.

7.8 Consider a WSS Gaussian process X[n] with autocorrelation function $R_X[k] = A\alpha^{|k|}$, where $A > 0$ and $0 < \alpha < 1$.

(a) Calculate the variance of the process.
(b) Determine the autocorrelation coefficient $\rho_X[k] = \frac{R_X[k]}{R_X[0]}$.

7.9 Stock prices are modeled as a Gaussian process. The prices over 5 days are: [102, 104, 107, 105, 108]. Assume a squared exponential kernel with parameters $\sigma_f^2 = 9$ and $\ell = 2$. Formulate the covariance matrix for these 5 days.

7.10 A discrete-time Gaussian random process X[n] has a power spectral density $S_X(e^{j\omega}) = 2 + 4\cos(\omega) + \cos(2\omega)$. Determine the autocorrelation function $R_X[k]$.

7.11 A WSS Gaussian random process X[n] has mean $\mu_X = 0$ and autocovariance function $C_X[k] = \sigma^2\delta[k]$, where $\delta[k]$ is the Kronecker delta function. If Y[n] = X[n] + X[n-1], find the autocovariance function $C_Y[k]$.

7.12 Two independent WSS Gaussian random processes X[n] and Y[n] have autocorrelation functions $R_X[k] = (0.5)^{|k|}$ and $R_Y[k] = 2\delta[k]$. If Z[n] = X[n] + Y[n], find the autocorrelation function $R_Z[k]$.

7.13 A discrete-time Gaussian random process X[n] is passed through a linear time-invariant (LTI) system with impulse response h[n] = 1, 0.5, 0.25. If the autocorrelation function of X[n] is $R_X[k] = \delta[k]$, find the autocorrelation function of the output Y[n].

7.14 A WSS Gaussian process X[n] has autocorrelation function $R_X[k] = 2(0.8)^{|k|}$. Find the power in the process X[n].

7.15 In a digital communication system, the received signal is $Y[n] = X[n] + W[n]$, where $X[n]$ is the transmitted signal with power $P_X = 8$, and $W[n]$ is AWGN with variance $\sigma^2 = 0.4$. Calculate the probability of error for BPSK modulation, given by $P_e = Q(\sqrt{\frac{2P_X}{\sigma^2}})$, where $Q(x)$ is the Q-function, and approximate $Q(6.32)$ as 10^{-9}.

7.16 A discrete-time Gaussian random vector $\mathbf{X} = [X_1, X_2]^T$ has a mean vector $\mu = [1, 2]^T$ and a covariance matrix $C = \begin{bmatrix} 2 & 1 \\ 1 & 3 \end{bmatrix}$. Find the correlation coefficient between X_1 and X_2.

7.17 A WSS Gaussian process X[n] has autocorrelation function $R_X[k] = 4\delta[k] + (0.6)^{|k|}$. Calculate the power spectral density $S_X(e^{j\omega})$ and sketch it for $\omega \in [-\pi, \pi]$.

7.18 A discrete-time Gaussian random process X[n] is used to model the height of waves in the sea. If the process has a mean height of 5 meters and an autocorrelation function that decays exponentially with lag, describe how you would estimate the average wave height over a long period.

7.19 In a tracking system using a Kalman filter, the measurement noise variance R is a critical parameter. Explain how an incorrect estimate of R can affect the filter's performance and the accuracy of the state estimates.

7.20 A financial model uses a Gaussian process to predict stock prices. Discuss how the choice of the covariance kernel (e.g., squared exponential, periodic) influences the model's ability to capture different patterns in stock price movements.

Chapter 8
Markov Processes

In this chapter, we discuss the foundational and advanced aspects of the Markov processes, a cornerstone of stochastic modeling and statistical signal processing. The simplicity and robustness of Markov processes arise from their unique *memoryless* property: the prediction of future states depends solely on the current state, independent of the sequence of events that led there. When the process evolves over a finite state space and discrete time intervals, it is referred to as a Discrete-Time Markov Chain (DTMC). This chapter introduces the mathematical underpinnings, classifications, dynamic behaviors, and real-world applications of Markov processes, particularly in the context of statistical signal processing.

8.1 Fundamentals of Markov Processes

A *Markov process* is a particular class of stochastic processes that plays a fundamental role in the mathematical modeling of random systems evolving over time. What distinguishes a Markov process from other types of stochastic processes is the Markov property, which asserts that the future evolution of the process is conditionally independent of its past, given the present state. This implies that the process is memoryless, which means its next state depends only on the current state and not on the full sequence of previous states.

Let $\{X_n\}_{n \geq 0}$ be a stochastic process, where X_n denotes the state of the process at a specific time n. The Markov property is formally defined as:

$$P(X_{n+1} = x_{n+1} \mid X_n = x_n, X_{n-1} = x_{n-1}, \ldots, X_0 = x_0) = P(X_{n+1} = x_{n+1} \mid X_n = x_n) \tag{8.1}$$

M. Catak and T. Allahviranloo, *Discrete Random Processes*, Studies in Systems, Decision and Control 644, https://doi.org/10.1007/978-3-032-08630-3_8

Equation (8.1) highlights that the conditional probability distribution of the future state X_{n+1}, given the entire history $X_0, X_1, \ldots, X_n$, is equal to the conditional probability given only the current state X_n. This property significantly reduces the complexity of analyzing and simulating stochastic systems, as it allows one to model the process using a sequence of transition probabilities that depend only on the present state.

Markov processes can be classified based on the nature of the state space and the time index set. When the state space is discrete and time progresses in discrete steps, the process is referred to as a discrete-time Markov chain. If either the state space or time is continuous, we refer to the process as a continuous-time Markov process or a Markov process with continuous state space, respectively.

The Markov property forms the theoretical foundation for numerous applications across a wide range of scientific and engineering disciplines. In statistical signal processing, for instance, Markov models are used to describe systems with evolving hidden states, such as in Hidden Markov Models (HMMs). In queueing theory, they provide tractable models for the analysis of service systems. In finance, Markov processes underpin models of asset price dynamics, such as the geometric Brownian motion model used in option pricing.

Moreover, due to their recursive structure, Markov processes lend themselves well to dynamic programming and optimal control strategies, where decisions at each time step depend only on the current state, aligning naturally with the memoryless structure. The recursive structure highlights how the future behavior of the Markov process can be understood by iteratively considering the transitions between adjacent states. This inherent structure makes Markov processes particularly well-suited for dynamic programming, where one can break down a complex problem into simpler, overlapping subproblems defined by the states of the process. The optimal decision at each step relies only on the current state, mirroring the recursive nature of how the process evolves.

8.1.1 Building Up a Markov Process

A Discrete-Time Markov Chain (DTMC) delineates the evolution of a system across a discrete temporal domain, transitioning between a finite set of discernible states. The defining characteristic of a Markov chain is its adherence to the Markov property, signifying that the probabilistic transition to a future state is exclusively contingent upon the current state, independent of the system's prior trajectory. The comprehensive specification of a DTMC necessitates the definition of three fundamental constituents: the initial state distribution, the transition probabilities governing state transitions, and a matrix representation of these probabilities.

Initial State Distribution ($\mathbf{a}[n]$)

The temporal evolution of the Markov process is initiated by the initial state distribution vector, denoted as $\mathbf{a}[n]$, which quantifies the probability of the system occupying each possible state at a specific initial time n. This vector is formally expressed as:

$$\mathbf{a}[n] = [a_1[n], a_2[n], \ldots, a_N[n]] \tag{8.2}$$

where $a_i[n]$ represents the probability measure of the system being in state i at time n, and N denotes the cardinality of the state space.
For $\mathbf{a}[n]$ to constitute a valid probability distribution, it must satisfy the following axiomatic conditions:

Normalization Condition: The sum of the probabilities across all possible states at any given time n must equal unity, reflecting the certainty that the system resides in one of the defined states.

$$\sum_{i=1}^{N} a_i[n] = 1 \tag{8.3}$$

Non-negativity Constraint: The probability of the system being in any specific state must be non-negative.

$$a_i[n] \geq 0 \quad \forall i \in \{1, 2, \ldots, N\} \tag{8.4}$$

Transition Probabilities (p_{ij})
The stochastic dynamics of the Markov process are governed by the transition probabilities, which define the likelihood of transitioning from a given state to another within a single discrete time step.
The transition probability p_{ij} is formally defined as the conditional probability of transitioning to state j at time $n+1$, given that the system was in state i at time n:

$$p_{ij} = P(X_{n+1} = j \mid X_n = i) \tag{8.5}$$

In accordance with the Markov property, this probability is solely a function of the current state i.
Transition Matrix (P)
The ensemble of all possible transition probabilities is systematically organized into the transition matrix, denoted by $\mathbf{P}$. This $N \times N$ matrix has elements p_{ij} at the i-th row and j-th column:

$$\mathbf{P} = \begin{bmatrix} p_{11} & p_{12} & \cdots & p_{1N} \\ p_{21} & p_{22} & \cdots & p_{2N} \\ \vdots & \vdots & \ddots & \vdots \\ p_{N1} & p_{N2} & \cdots & p_{NN} \end{bmatrix} \tag{8.6}$$

A fundamental property of the transition matrix is that the elements of each row sum to unity, reflecting the requirement that from any given state i, the system must transition to one of the N possible states with probability 1. Furthermore, each individual transition probability is bounded between 0 and 1, inclusive.

$$\sum_{j=1}^{N} p_{ij} = 1, \quad 0 \leq p_{ij} \leq 1 \quad \forall i, j \in \{1, 2, \ldots, N\} \tag{8.7}$$

Temporal Evolution of the State Probability Vector:
The temporal evolution of the state probability vector is determined by the transition matrix. Given the state probability vector at time n, $\mathbf{a}[n]$, the state probability vector at the subsequent time step, $\mathbf{a}[n+1]$, can be computed through matrix multiplication:

$$\mathbf{a}[n+1] = \mathbf{a}[n]\mathbf{P} \tag{8.8}$$

This equation explains how the probabilistic distribution over the state space evolves over time, with the probabilities at each time step being a function of the probabilities at the preceding time step and the defined transition dynamics. Iterative application of this relationship allows for the probabilistic forecasting of the system's state at any future time.

8.2 Classification of States in Markov Chains

The states within a Markov chain exhibit diverse behaviors and properties that are crucial for understanding the long-term dynamics of the system. These classifications provide a framework for analyzing and predicting the evolution of the chain. We will delve into each category with detailed explanations and illustrative examples.

Accessible State:
A state j is deemed **accessible** from state i if there exists a non-negative integer $n \geq 0$ such that the probability of being in state j at time n, given that the chain started in state i at time 0, is strictly greater than zero. Mathematically, this is expressed as $P(X_n = j \mid X_0 = i) > 0$. This implies that there is at least one sequence of transitions of length n that allows the system to move from state i to state j. The number of steps n can vary depending on the specific pair of states.
Consider a Markov chain with three states $\{1, 2, 3\}$ and the following transition matrix:

$$\mathbf{P} = \begin{bmatrix} 0.5 & 0.5 & 0 \\ 0 & 0.3 & 0.7 \\ 0.8 & 0.2 & 0 \end{bmatrix}$$

From state 1, state 2 is accessible in one step ($p_{12} = 0.5 > 0$). State 3 is not accessible in one step ($p_{13} = 0$). However, state 3 is accessible in two steps (e.g., $1 \to 2 \to 3$ with probability $0.5 \times 0.7 = 0.35 > 0$). State 1 is accessible from state 3 in one step ($p_{31} = 0.8 > 0$).

Communication:
Two states, i and j, are said to **communicate** if state j is accessible from state i, *and* state i is accessible from state j. This signifies a mutual reachability between the two states. We denote this relationship as $i \leftrightarrow j$. Communication is an equivalence relation, meaning it is reflexive ($i \leftrightarrow i$), symmetric (if $i \leftrightarrow j$, then $j \leftrightarrow i$), and transitive (if $i \leftrightarrow j$ and $j \leftrightarrow k$, then $i \leftrightarrow k$).
Considering the same transition matrix given above:

$$\mathbf{P} = \begin{bmatrix} 0.5 & 0.5 & 0 \\ 0 & 0.3 & 0.7 \\ 0.8 & 0.2 & 0 \end{bmatrix}$$

State 1 and state 2 communicate because state 2 is accessible from state 1 (in 1 step), and state 1 is accessible from state 2 (e.g., $2 \to 3 \to 1$ with probability $0.7 \times 0.8 = 0.56 > 0$).
State 2 and state 3 communicate because state 3 is accessible from state 2 (in 1 step), and state 2 is accessible from state 3 (in 1 step).
Due to transitivity, state 1 and state 3 also communicate.

Irreducibility:
A Markov chain is called **irreducible** if all states within the chain communicate with each other. In an irreducible chain, it is possible to reach any state from any other state, not necessarily in a single step. Irreducibility implies that the state space consists of a single communicating class.
The Markov chain with the transition matrix:

$$\mathbf{P} = \begin{bmatrix} 0.5 & 0.5 & 0 \\ 0 & 0.3 & 0.7 \\ 0.8 & 0.2 & 0 \end{bmatrix}$$

is irreducible because we established that states 1 and 2 communicate, states 2 and 3 communicate, and states 1 and 3 communicate.
On the other hand, consider a Markov chain with three states and the transition matrix:

$$\mathbf{P} = \begin{bmatrix} 1 & 0 & 0 \\ 0.2 & 0.6 & 0.2 \\ 0 & 0.5 & 0.5 \end{bmatrix}$$

State 1 is an absorbing state. States 2 and 3 communicate with each other. However, it is impossible to reach state 2 or state 3 from state 1. Therefore, this chain is reducible, with communicating states $\{1\}$ and $\{2, 3\}$.

Transient State:
A state i is **transient** if there is a non-zero probability that the chain will never return to state i after leaving it. Formally, $P(\text{return to state } i \mid X_0 = i) < 1$. In a Markov chain with a finite number of states, at least one state must be recurrent. Transient states are visited only a finite number of times with probability 1.
Using the reducible chain matrix:

$$\mathbf{P} = \begin{bmatrix} 1 & 0 & 0 \\ 0.2 & 0.6 & 0.2 \\ 0 & 0.5 & 0.5 \end{bmatrix}$$

States 2 and 3 are transient. Once the chain enters state 1, it can never leave. Therefore, there is a non-zero probability (in fact, a probability of 0 for state 2 and 3 if it starts in state 1) that the chain will not return to states 2 or 3 after potentially visiting them.

Recurrent State:
A state i is **recurrent** if, starting from state i, the probability of eventually returning to state i is 1. Formally, $P(\text{return to state } i \mid X_0 = i) = 1$. In a finite-state Markov chain, if a state is not transient, it must be recurrent. Recurrent states are visited infinitely often with probability 1.
Using the reducible chain matrix:

$$\mathbf{P} = \begin{bmatrix} 1 & 0 & 0 \\ 0.2 & 0.6 & 0.2 \\ 0 & 0.5 & 0.5 \end{bmatrix}$$

State 1 is a recurrent state (specifically, an absorbing state, which is a type of recurrent state). Once the chain enters state 1, it stays there with probability 1.

Absorbing State:
An **absorbing state** i is a special type of recurrent state where, once the chain enters this state, it can never leave. This is characterized by a transition probability $p_{ii} = 1$ and $p_{ij} = 0$ for all $j \neq i$.
In the matrix:

$$\mathbf{P} = \begin{bmatrix} 1 & 0 & 0 \\ 0.2 & 0.6 & 0.2 \\ 0 & 0.5 & 0.5 \end{bmatrix}$$

State 1 is an absorbing state.

Periodic State:
A state i is **periodic** with period $k > 1$ if any return to state i must occur in a number of steps that is a multiple of k, and k is the largest such integer. Formally, $P(X_n = i \mid X_0 = i) > 0$ only if n is a multiple of k, and $k = \gcd\{n > 0 \mid P(X_n = i \mid X_0 = i) > 0\}$. If $k = 1$, the state is aperiodic. Periodic states exhibit cyclical behavior.
Consider a Markov chain with three states and the transition matrix:

$$\mathbf{P} = \begin{bmatrix} 0 & 1 & 0 \\ 0 & 0 & 1 \\ 1 & 0 & 0 \end{bmatrix}$$

Starting from state 1, the chain can only return to state 1 in 3 steps ($1 \rightarrow 2 \rightarrow 3 \rightarrow 1$), or multiples of 3 (e.g., $1 \rightarrow 2 \rightarrow 3 \rightarrow 1 \rightarrow 2 \rightarrow 3 \rightarrow 1$ in 6 steps). The period of state 1 is 3. Similarly, states 2 and 3 also have a period of 3.
Another periodic case can be implemented using the following transition matrix:

$$\mathbf{P} = \begin{bmatrix} 0 & 1 & 0 & 0 \\ 1 & 0 & 0 & 0 \\ 0 & 0 & 0.5 & 0.5 \\ 0 & 0 & 0.5 & 0.5 \end{bmatrix}$$

States 1 and 2 have a period of 2 ($1 \leftrightarrow 2$). States 3 and 4 are aperiodic (there is a self-loop).

Ergodic State:
A state i is **ergodic** if it is both positive recurrent and aperiodic.

> **Positive Recurrence:** A recurrent state i is positive recurrent if the expected return time to state i, starting from i, is finite.
> **Aperiodicity:** A state i is aperiodic if the greatest common divisor (gcd) of all possible return times to state i is 1. In simpler terms, there is no fixed cycle in which the state must be revisited.

Ergodic states exhibit stable long-term behavior. In an irreducible Markov chain where all states are ergodic, the chain is called an ergodic Markov chain, and it has a unique stationary distribution.
Consider an irreducible Markov chain with the transition matrix:

$$\mathbf{P} = \begin{bmatrix} 0 & 1 \\ 0.5 & 0.5 \end{bmatrix}$$

Both states are recurrent. To check for periodicity, consider state 1. Possible return times are 2 ($1 \rightarrow 2 \rightarrow 1$), 3 ($1 \rightarrow 2 \rightarrow 2 \rightarrow 1$ with probability 0, so consider longer paths: $1 \rightarrow 2 \rightarrow 1 \rightarrow 2 \rightarrow 1$ in 4 steps, etc.). The gcd of the possible return times is 1, so state 1 is aperiodic. Similarly, state 2 is also aperiodic. If we assume positive recurrence (which can be shown), then both states are ergodic.

8.3 Random Walk Example

8.3.1 One-Dimensional Symmetric Random Walk

Consider a simple symmetric random walk on the integer line. At each time step, the walker moves:

- One unit to the right with probability $p = 0.5$
- One unit to the left with probability $q = 0.5$

Let X_n denote the position at time n. The transition probabilities are:

$$P(X_{n+1} = i + 1 \mid X_n = i) = 0.5, \quad P(X_{n+1} = i - 1 \mid X_n = i) = 0.5 \tag{8.9}$$

The expected position after n steps is $E[X_n] = 0$, and the variance is $Var(X_n) = n$. The probability of returning to the origin infinitely many times is 1.

A realization starting from $X_0 = 0$ over 10 time steps might be:

$$(0, 1, 0, -1, -2, -1, -2, -1, 0, 1, 2)$$

An illustrative realization of a symmetric random walk is given in Fig. 8.1.

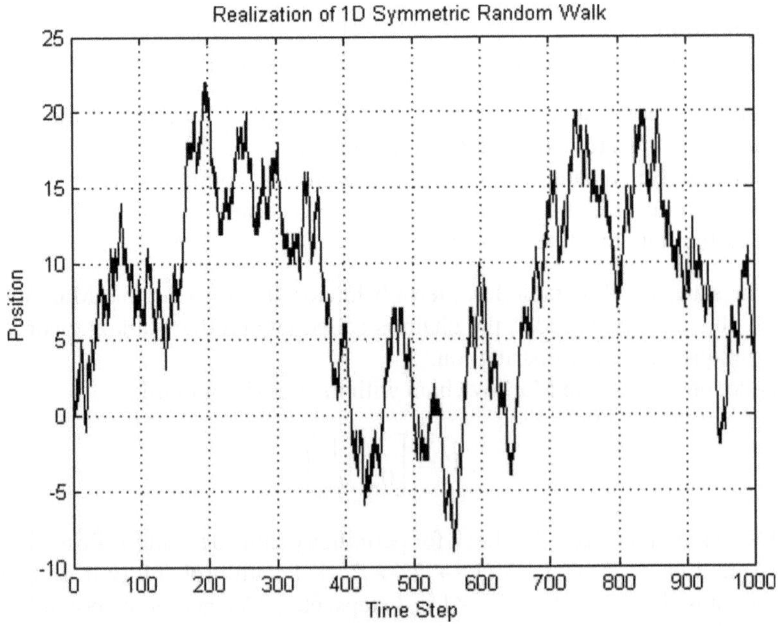

Fig. 8.1 A realization of a symmetric random walk

In Fig. 8.1, horizontal axis represents the progression of time, marked in discrete steps from 0 to 1000. Each integer value on this axis corresponds to an instance in the walk. Vertical axis, shows the position of the walker on the one-dimensional line at each corresponding time step. The position starts at 0 as an initial condition and fluctuates over time, ranging from -6 to 23 in this particular realization. The line plotted on the graph traces the path of the random walker over these 1000 time steps. Each small segment of the line represents a single step taken by the walker. The upward segments indicate a step to the right (increasing position), and the downward segments indicate a step to the left (decreasing position). There's no clear trend or pattern in the movement, which is characteristic of a random process. A fundamental property of a simple random walk is that each step is independent of the previous steps. The direction of the next step is not influenced by where the walker has been. This "lack of memory" contributes to the irregular path.

While this figure only shows 1000 steps, one can infer some long-term behaviors of random walks. A symmetric random walk is recurrent, meaning that the walker will, with probability 1, return to its starting position infinitely many times as the number of steps approaches infinity. While this specific realization might not be at 0 at the end, given enough time, it's highly likely to revisit it.

8.3.2 One-Dimensional Nonsymmetric Random Walk

Consider a nonsymmetric random walk where the probabilities of moving right and left are different:

- One unit to the right with probability p
- One unit to the left with probability $q = 1 - p$, where $p \neq 0.5$

The transition probabilities are:

$$P(X_{n+1} = i + 1 \mid X_n = i) = p, \quad P(X_{n+1} = i - 1 \mid X_n = i) = 1 - p \tag{8.10}$$

The expected position after n steps is $E[X_n] = n(2p - 1)$, and the variance is $Var(X_n) = 4np(1 - p)$. The probability of returning to the origin infinitely many times is less than 1.

An realization of a nonsymmetric random walk is shown in Fig. 8.2.

The significant difference here is the term nonsymmetric and the introduction of the probability value $p = 0.55$. It means the probability of moving in one direction, typically considered the positive direction or right, is 0.55 while the probability of moving in the opposite direction, negative or left, is $1 - p = 0.45$. These probabilities are no longer equal, introducing a bias into the walk. In Fig. 8.2, the plotted line shows the cumulative position of the walker over the 1000 time steps. Noticeably, there is a clear upward trend in this path, unlike the more oscillatory behavior of the symmetric random walk.

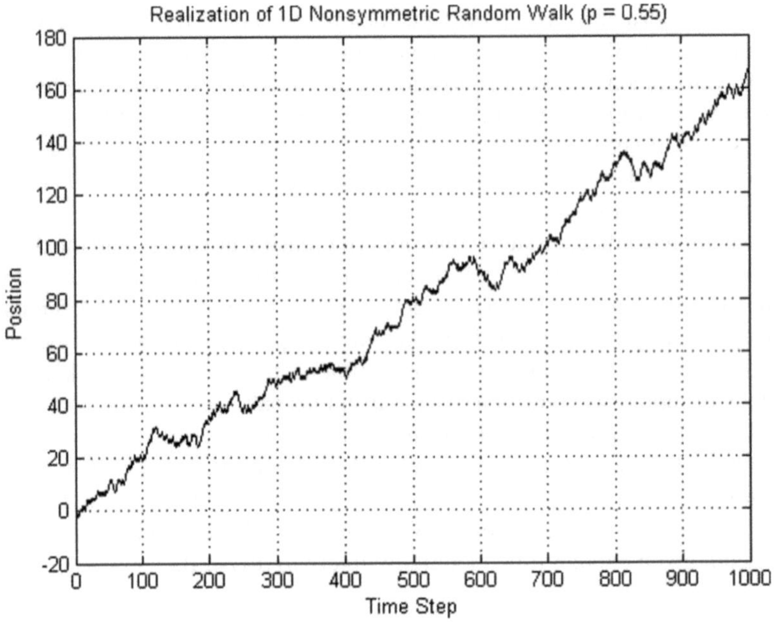

Fig. 8.2 A realization of a nonsymmetric random walk

8.4 Long-Term Behavior and Limiting Distribution

For a discrete-time Markov chain with a finite state space, the long-term behavior of the system can often be described by its limiting distribution. If the Markov chain possesses certain properties, specifically irreducibility and aperiodicity, it will converge to a unique stationary distribution.

As a reminder, *reducibility* means that it is possible to reach any state from any other state in a finite number of steps. There are no closed sets of states that the chain can get trapped in. *Aperiodicity* means that the chain does not get trapped in cycles of states. For any state i, the greatest common divisor of all possible return times to state i is 1.

If a Markov chain is irreducible and aperiodic, then a unique *stationary distribution* $\boldsymbol{\pi} = [\pi_1, \pi_2, \ldots, \pi_m]$ exists, where $\pi_i \geq 0$ for all states i, and $\sum_i \pi_i = 1$. This stationary distribution satisfies the following equation:

$$\boldsymbol{\pi}\mathbf{P} = \boldsymbol{\pi}, \quad \sum_i \pi_i = 1 \tag{8.11}$$

Here, the equation $\boldsymbol{\pi}\mathbf{P} = \boldsymbol{\pi}$ implies that once the distribution of states reaches $\boldsymbol{\pi}$, it remains unchanged over time.

As the number of time steps n approaches infinity, the state distribution vector $\mathbf{a}[n] = [a_1[n], a_2[n], \ldots, a_m[n]]$, where $a_i[n]$ is the probability of being in state i at time n, converges to the stationary distribution $\boldsymbol{\pi}$, regardless of the initial distribution $\mathbf{a}[0]$:

$$\mathbf{a}[n] = \mathbf{a}[0]\mathbf{P}^n \to \boldsymbol{\pi} \quad \text{as } n \to \infty \tag{8.12}$$

This convergence implies that in the long run, the probability of being in each state becomes time-invariant and is given by the components of the stationary distribution $\boldsymbol{\pi}$.

Consider a simple Markov chain with three states: State 1, State 2, and State 3. Let the transition matrix $\mathbf{P}$ be:

$$\mathbf{P} = \begin{bmatrix} 0.5 & 0.3 & 0.2 \\ 0.4 & 0.4 & 0.2 \\ 0.1 & 0.5 & 0.4 \end{bmatrix}$$

Here, $P_{12} = 0.3$ means there is a 30% probability of transitioning from State 1 to State 2 in one step. We can observe that all entries are non-negative and each row sums to 1, which are the basic requirements for a transition matrix. Let's assume this Markov chain is irreducible and aperiodic.

To find the long-term distribution $\boldsymbol{\pi} = [\pi_1, \pi_2, \pi_3]$, we need to solve the system of linear equations $\boldsymbol{\pi}\mathbf{P} = \boldsymbol{\pi}$ and $\sum_i \pi_i = 1$. This can be written as:

$$\begin{aligned} \pi_1(0.5) + \pi_2(0.4) + \pi_3(0.1) &= \pi_1 \\ \pi_1(0.3) + \pi_2(0.4) + \pi_3(0.5) &= \pi_2 \\ \pi_1(0.2) + \pi_2(0.2) + \pi_3(0.4) &= \pi_3 \\ \pi_1 + \pi_2 + \pi_3 &= 1 \end{aligned}$$

Rearranging the first three equations:

$$\begin{aligned} -0.5\pi_1 + 0.4\pi_2 + 0.1\pi_3 &= 0 \\ 0.3\pi_1 - 0.6\pi_2 + 0.5\pi_3 &= 0 \\ 0.2\pi_1 + 0.2\pi_2 - 0.6\pi_3 &= 0 \\ \pi_1 + \pi_2 + \pi_3 &= 1 \end{aligned}$$

Solving this system of linear equations, the long-term distribution is approximately:

$$\pi = \left[\frac{13}{36}, \frac{7}{18}, \frac{1}{4}\right] \approx [0.3611, 0.3889, 0.25]$$

This means that in the long run, the probability of being in State 1 is approximately 0.3611, the probability of being in State 2 is approximately 0.3889, and the probability of being in State 3 is 0.25, regardless of the initial state distribution.

8.4.1 *Eigenvalues of the Transition Matrix and System Analysis*

The eigenvalues and eigenvectors of the transition matrix $\mathbf{P}$ are powerful tools for analyzing the dynamic behavior of a Markov chain. They provide insights into the long-term stability, the rate of convergence to a steady state, and the fundamental modes of evolution of the system.

Significance of Eigenvalues:

The eigenvalues λ_i of the transition matrix $\mathbf{P}$ are crucial because they characterize the scaling factors of the eigenvectors under the linear transformation represented by $\mathbf{P}$. When we apply the transition matrix $\mathbf{P}$ to the state probability vector $\mathbf{a}[n]$, we are essentially projecting this vector onto the eigenvectors and scaling each component by its corresponding eigenvalue.

Eigenvalue of 1 and Stationary Distribution:

A fundamental property of a stochastic matrix (a transition matrix where rows sum to 1 and elements are non-negative) is that it always has at least one eigenvalue equal to 1. If the Markov chain is irreducible and aperiodic, which means ergodic, this eigenvalue of 1 is unique (non-degenerate) among eigenvalues with magnitude 1. The left eigenvector associated with this eigenvalue of 1 is the stationary distribution $\boldsymbol{\pi}$. This is because if $\boldsymbol{\pi}$ is the stationary distribution, it satisfies $\boldsymbol{\pi}\mathbf{P} = \boldsymbol{\pi}$, which can be rewritten as $\boldsymbol{\pi}\mathbf{P} = 1 \cdot \boldsymbol{\pi}$, indicating that $\boldsymbol{\pi}$ is a left eigenvector of $\mathbf{P}$ with eigenvalue 1.

Rate of Convergence:

The magnitudes of the other eigenvalues (those not equal to 1) determine the rate at which the Markov chain converges to its stationary distribution. Let the eigenvalues be ordered such that $|\lambda_1| = 1 \geq |\lambda_2| \geq |\lambda_3| \geq \cdots \geq |\lambda_N|$. The second largest eigenvalue in magnitude, $|\lambda_2|$, plays a critical role in determining the convergence rate. The closer $|\lambda_2|$ is to 1, the slower the convergence; the smaller $|\lambda_2|$ is, the faster the convergence. If all other eigenvalues have a magnitude strictly less than 1 ($|\lambda_i| < 1$ for $i > 1$), the chain will eventually converge to the stationary distribution, regardless of the initial distribution.

Stability:

Since all eigenvalues of a stochastic matrix have an absolute value less than or equal to 1, the powers of the transition matrix $\mathbf{P}^n$ remain bounded as $n \to \infty$. This ensures that the probabilities in the state distribution vector $\mathbf{a}[n] = \mathbf{a}[0]\mathbf{P}^n$ also remain bounded between 0 and 1, which is a necessary condition for a probabilistic system.

Significance of Eigenvectors:

The eigenvectors associated with the eigenvalues provide the fundamental modes or components of the system's evolution.

Eigenvector for Eigenvalue 1:

As mentioned earlier, the left eigenvector corresponding to the eigenvalue $\lambda = 1$ is the stationary distribution $\boldsymbol{\pi}$. This vector represents the long-term probabilities of being in each state.

Other Eigenvectors and Transient Behavior:

The other eigenvectors, associated with eigenvalues $|\lambda_i| < 1$ for $i > 1$, represent the transient components of the state distribution. These components decay over time as we take higher powers of $\mathbf{P}$ ($\mathbf{P}^n$), because $\lambda_i^n \to 0$ as $n \to \infty$ for $|\lambda_i| < 1$. The initial state distribution can be expressed as a linear combination of all the left eigenvectors. As time progresses, the components along the eigenvectors with magnitudes less than 1 diminish, leaving only the component along the eigenvector with eigenvalue 1, which is the stationary distribution.

A Simple 2-State Markov Chain
Consider a Markov chain with two states and the transition matrix:

$$\mathbf{P} = \begin{bmatrix} 0.8 & 0.2 \\ 0.3 & 0.7 \end{bmatrix}$$

The characteristic equation is

$$\det(\mathbf{P} - \lambda\mathbf{I}) = \det \begin{bmatrix} 0.8 - \lambda & 0.2 \\ 0.3 & 0.7 - \lambda \end{bmatrix} = (0.8 - \lambda)(0.7 - \lambda) - (0.2)(0.3) = 0$$

Expanding this gives

$$\lambda^2 - 1.5\lambda + 0.56 - 0.06 = \lambda^2 - 1.5\lambda + 0.5 = (\lambda - 1)(\lambda - 0.5) = 0$$

The eigenvalues are $\lambda_1 = 1$ and $\lambda_2 = 0.5$.

Eigenvalue $\lambda_1 = 1$:

To find the left eigenvector $\boldsymbol{\pi} = [\pi_1, \pi_2]$, we solve $\boldsymbol{\pi}\mathbf{P} = \boldsymbol{\pi}$, which gives:

$$0.8\pi_1 + 0.3\pi_2 = \pi_1 \quad \Rightarrow \quad -0.2\pi_1 + 0.3\pi_2 = 0$$
$$0.2\pi_1 + 0.7\pi_2 = \pi_2 \quad \Rightarrow \quad 0.2\pi_1 - 0.3\pi_2 = 0$$

And the normalization condition $\pi_1 + \pi_2 = 1$. Solving these, we get $\pi_1 = 0.6$ and $\pi_2 = 0.4$. So, the stationary distribution is $\boldsymbol{\pi} = [0.6, 0.4]$.

Eigenvalue $\lambda_2 = 0.5$:

The magnitude of this eigenvalue is $|0.5| < 1$, indicating convergence. The rate of convergence is governed by this value.

A 3-State Markov Chain
Consider the transition matrix:

$$\mathbf{P} = \begin{bmatrix} 0.5 & 0.3 & 0.2 \\ 0.4 & 0.4 & 0.2 \\ 0.1 & 0.5 & 0.4 \end{bmatrix}$$

The characteristic equation $\det(\mathbf{P} - \lambda \mathbf{I}) = 0$ will be a cubic equation in λ. Solving this yields three eigenvalues: $\lambda_1 = 1$, and two others λ_2, λ_3 with magnitudes less than or equal to 1. (The exact values are approximately $\lambda_1 = 1$, $\lambda_2 \approx 0.2 + 0.3464i$, $\lambda_3 \approx 0.2 - 0.3464i$.)

Eigenvalue $\lambda_1 = 1$:

The left eigenvector corresponding to $\lambda_1 = 1$ is the stationary distribution $\boldsymbol{\pi} \approx [0.3611, 0.3889, 0.25]$.

Eigenvalues $\lambda_2 \approx 0.2 + 0.3464i$, $\lambda_3 \approx 0.2 - 0.3464i$:

The magnitudes of these eigenvalues are

$$|\lambda_2| = |\lambda_3| = \sqrt{0.2^2 + 0.3464^2} = \sqrt{0.04 + 0.12} = \sqrt{0.16} = 0.4$$

Since $|0.4| < 1$, the system will converge to the stationary distribution. The fact that they are complex indicates that the convergence might involve some oscillatory behavior. The rate of decay of the transient components is governed by the magnitude 0.4.

If the transition matrix $\mathbf{P}$ is diagonalizable, it can be expressed as $\mathbf{P} = \mathbf{E}\mathbf{D}\mathbf{E}^{-1}$, where $\mathbf{D}$ is a diagonal matrix containing the eigenvalues.

8.5 Applications of Markov Processes in Statistical Signal Processing

8.5.1 Queuing Theory: Modeling a Simple Queue

Queuing theory is the mathematical study of waiting lines (or queues). It is significant in electrical engineering for analyzing network traffic, telecommunication systems, and computer performance. Markov chains provide a powerful tool to model the dynamic behavior of queues.

Let us consider a simple queuing system with a single server. Customers arrive at the system, and if the server is busy, they join a queue. The following assumptions are supposed to model such system by means of Markov chains:

- Discrete Time:We observe the system at discrete time intervals (e.g., every second).
- State Definition: The state of the system, denoted by X_n, is the number of customers in the system at time n (including those being served and those waiting). The state space is $S = \{0, 1, 2, 3, ...\}$.
- Markov Property: The number of customers at time $n + 1$ depends only on the number of customers at time n, and not on the history of the system.
- Arrivals and Departures: In each time interval, there's a probability p that a new customer arrives and a probability q that a customer being served departs (if there's a customer being served). It is assumed these events are statistically independent.

The transition probabilities between the states can be defined as:
$p_{i,i+1} = p(1 - q)$ for $i \geq 0$ (One arrival, no departure)
$p_{i,i-1} = q(1 - p)$ for $i \geq 1$ (One departure, no arrival)
$p_{i,i} = pq + (1 - p)(1 - q)$ for $i \geq 1$ (One arrival and one departure, or no arrival and no departure)
$p_{0,0} = (1 - p) + pq$ (No arrival, or one arrival and one departure)
$p_{i,j} = 0$ otherwise

This means the system evolves as a Markov chain, and one can analyze properties like: i) The probability of having a certain number of customers in the system in the long run, ii) The average queue length. iii) The average waiting time.

Suppose one observe a network router's queue every millisecond. The probability of a new packet arriving in a millisecond is $p = 0.1$, and the probability of a packet being processed and leaving in a millisecond is $q = 0.2$.

The transition matrix **P** for this system (for the first few states) would be:

$$\mathbf{P} = \begin{bmatrix} 0.9+0.02 & 0.1(0.8) & 0 & 0 & 0 & \cdots \\ 0.2(0.9) & 0.02+0.88 & 0.08 & 0 & 0 & \cdots \\ 0 & 0.18 & 0.02+0.88 & 0.08 & 0 & \cdots \\ 0 & 0 & 0.18 & 0.9 & 0.08 & \cdots \\ 0 & 0 & 0 & 0.18 & 0.9 & \cdots \\ \vdots & \vdots & \vdots & \vdots & \vdots & \ddots \end{bmatrix} = \begin{bmatrix} 0.92 & 0.08 & 0 & 0 & 0 & \cdots \\ 0.18 & 0.9 & 0.08 & 0 & 0 & \cdots \\ 0 & 0.18 & 0.9 & 0.08 & 0 & \cdots \\ 0 & 0 & 0.18 & 0.9 & 0.08 & \cdots \\ 0 & 0 & 0 & 0.18 & 0.9 & \cdots \\ \vdots & \vdots & \vdots & \vdots & \vdots & \ddots \end{bmatrix}$$

Using this transition matrix, wone can, for example, calculate the probability of having 2 packets in the queue after 5 milliseconds, given that we started with 0 packets. One could also find the stationary distribution to determine the long-term average number of packets in the queue. The system parameters can be listed as:

- Probability of packet arrival per millisecond: $p = 0.1$
- Probability of packet departure per millisecond (if serving): $q = 0.2$
- Utilization factor: $\rho = \frac{p}{q} = 0.5$

The stationary distribution $\boldsymbol{\pi} = [\pi_0, \pi_1, \pi_2, ...]$ gives the long-run probabilities of having k packets in the system. For this $M/M/1$-like queue, it is given by:

$$\pi_k = (1 - \rho)\rho^k = (0.5)(0.5)^k = (0.5)^{k+1}$$

For instance, $\pi_0 = 0.5$, which means 50% probability of the system being empty; $\pi_1 = 0.25$, 25% probability of having only 1 packet; $\pi_2 = 0.125$, 12.5% probability of having 2 packets.

The average number of packets in the system can be derived as;

$$L = \frac{\rho}{1 - \rho} = \frac{0.5}{1 - 0.5} = 1 \text{ packet}$$

This means that, on average, there is one packet in the system at any given time.

The average time, W, a packet spends in the system is:

$$W = \frac{L}{\lambda} = \frac{1}{0.1} = 10 \text{ milliseconds}$$

where $\lambda = p$ is the arrival rate,

Therefore, a packet spends an average of 10 milliseconds in the system, including waiting and service time.

8.5.2 *Image Processing: Markov Random Fields for Image Denoising*

Markov Random Fields (MRFs) are a powerful tool in image processing for tasks like denoising, segmentation, and restoration. They extend the concept of Markov chains to multiple dimensions.

In image denoising, the goal is to recover the original, clean image from a noisy version. MRFs can be used to model the statistical dependencies between neighboring pixels. The following assumptions have been carried out:

- An image is represented as a grid of pixels. Each pixel has a value (e.g., intensity or color).
- Each pixel's value is considered a random variable.

- A neighborhood system is defined, where each pixel has a set of neighboring pixels (e.g., the 4 pixels directly above, below, left, and right).
- The key idea is the Markov property: the value of a pixel, given the values of all other pixels, depends only on the values of its neighbors. This captures the intuition that neighboring pixels tend to have similar values in a typical image.

It can be defined an energy function that captures these relationships:

$$E(x) = \sum_i \alpha (x_i - y_i)^2 + \beta \sum_{i \sim j} (x_i - x_j)^2$$

where: x is the restored image, and x_i is the restored value of pixel i; y is the noisy image, and y_i is the noisy value of pixel i; $i \sim j$ means that pixels i and j are neighbors; α and β are parameters that control the influence of the noisy data and the smoothness of the restored image. The first term encourages the restored image to be close to the noisy image, while the second term encourages neighboring pixels to have similar values (smoothness).

The denoising problem then becomes finding the image x that minimizes this energy function.

Consider a small 3×3 grayscale image with added Gaussian noise. The noisy image (y) is:

$$y = \begin{bmatrix} 102 & 110 & 108 \\ 105 & 98 & 101 \\ 100 & 103 & 99 \end{bmatrix}$$

Let us use a simple 4-neighborhood system and set $\alpha = 1$ and $\beta = 0.5$.

To denoise the image, each pixel's value will be iteratively updated based on its neighbors and the noisy data. After several iterations, the denoised image (x) will be converged as:

$$x = \begin{bmatrix} 104 & 109 & 107 \\ 104 & 100 & 102 \\ 101 & 102 & 100 \end{bmatrix}$$

Problems

8.1 Consider a Markov chain with three states, $S = \{1, 2, 3\}$, and the following transition matrix:

$$\mathbf{P} = \begin{bmatrix} 0.6 & 0.2 & 0.2 \\ 0.3 & 0.5 & 0.2 \\ 0.1 & 0.4 & 0.5 \end{bmatrix}$$

(a) Draw the state transition diagram for this Markov chain.
(b) Determine if this Markov chain is:

 (i) Irreducible
 (ii) Aperiodic

 Justify your answers.
(c) Based on your answers to (b), is this Markov chain ergodic? Explain.
(d) Calculate $\mathbf{P}^2$ and $\mathbf{P}^3$. What do these matrices represent?
(e) Find the stationary distribution $\boldsymbol{\pi}$ of this Markov chain. Show your calculations.
(f) If the initial state distribution is $\mathbf{a}[0] = [0.5, 0.3, 0.2]$, calculate the state distribution after 2 steps, $\mathbf{a}[2]$.
(g) Does the limiting distribution of this Markov chain exist? If so, what is it?

8.2 Consider the following transition matrix for a Markov chain with states $S = \{A, B, C, D\}$:

$$\mathbf{P} = \begin{bmatrix} 0 & 1 & 0 & 0 \\ 0 & 0 & 1 & 0 \\ 0 & 0 & 0 & 1 \\ 1 & 0 & 0 & 0 \end{bmatrix}$$

(a) Draw the state transition diagram.
(b) Determine the period of each state. Is this chain periodic?
(c) Is this chain irreducible? Explain.
(d) Are the states in this chain recurrent or transient? Justify your answer.
(e) Does a limiting distribution exist for this chain? If so, find it. If not, explain why.

8.3 A Markov chain has the following transition matrix:

$$\mathbf{P} = \begin{bmatrix} 1/2 & 1/2 & 0 & 0 \\ 1/4 & 3/4 & 0 & 0 \\ 0 & 0 & 1/3 & 2/3 \\ 0 & 0 & 1/2 & 1/2 \end{bmatrix}$$

(a) Identify the recurrent and transient states in this chain.
(b) Do any absorbing states exist? If so, identify them.
(c) Does a stationary distribution exist for this chain?

8.4 Consider a particle performs a random walk on a one-dimensional integer lattice. At each time step, it moves one unit to the right with probability p and one unit to the left with probability $1 - p$. The particle starts at position 0 at time $n = 0$. For the case of a symmetric random walk, where $p = 0.5$:

(a) Calculate the probability that the particle is at position 2 after 3 time steps.
(b) What is the expected position of the particle after n time steps?
(c) Discuss the long-term behavior of the particle's position. Will it always return to the origin? Explain.

8.5 For the case of a nonsymmetric random walk, where $p = 0.6$:

(a) Calculate the probability that the particle is at position 2 after 3 time steps.
(b) What is the expected position of the particle after n time steps?
(c) Discuss the long-term behavior of the particle's position. Will it always return to the origin? Explain.

8.6 A simplified model of a network router's buffer as a queuing system is examined. Time is divided into discrete slots. In each slot, a new packet arrives with probability p, and a packet is transmitted, if there is one to transmit, with probability q. The buffer can hold an unlimited number of packets. Let X_n be the number of packets in the buffer at the beginning of time slot n.

(a) Draw the state transition diagram for this Markov chain.
(b) Write down the transition matrix **P** for this Markov chain (for the first few states).
Considering $p = 0.2$ and $q = 0.3$:
(c) Calculate the probability that there are 2 packets in the buffer at the beginning of time slot 3, given that there was 0 packet at the beginning of time slot 0.
(d) Find the stationary distribution of the number of packets in the buffer.
(e) Calculate the average number of packets in the buffer in the long run.
(f) Calculate the average time a packet spends in the buffer (in time slots).

8.7 A simplified model of a computer's CPU state transitions is as follows: It can be in one of three states: Idle (I), Busy (B), or Waiting for I/O (W). The transition matrix is given by:

$$\mathbf{P} = \begin{bmatrix} 0.2 & 0.6 & 0.2 \\ 0.3 & 0.5 & 0.2 \\ 0.4 & 0.2 & 0.4 \end{bmatrix}$$

(a) Draw the state transition diagram for this Markov chain.
(b) If the CPU starts in the Idle state, what is the probability it will be in the Busy state after 2 transitions?
(c) Determine the stationary distribution of the CPU states.
(d) Is this Markov chain ergodic? Justify your answer.

8.8 Consider a Markov chain with states $S = \{0, 1, 2\}$ and transition matrix:

$$\mathbf{P} = \begin{bmatrix} 0.5 & 0.5 & 0 \\ 0.2 & 0.6 & 0.2 \\ 0 & 0.4 & 0.6 \end{bmatrix}$$

(a) Draw the state transition diagram.
(b) Calculate $\mathbf{P}^2$.
(c) Find the stationary distribution of this chain.
(d) Is this chain irreducible and aperiodic?

8.9 A communication channel transmits binary data. Due to noise, a '0' can be received as a '1' with probability p, and a '1' can be received as a '0' with probability q. Model this as a Markov chain with two states: '0' and '1'.

(a) Write down the transition matrix.
(b) Draw the state transition diagram.
(c) Find the stationary distribution of this chain.

8.10 A simplified model of a stock price has three states: Increase (I), Stay the Same (S), and Decrease (D). The transition matrix is:

$$\mathbf{P} = \begin{bmatrix} 0.6 & 0.2 & 0.2 \\ 0.1 & 0.8 & 0.1 \\ 0.2 & 0.3 & 0.5 \end{bmatrix}$$

(a) Draw the state transition diagram.
(b) If the stock price is currently Increasing, what is the probability it will Decrease after 2 periods?
(c) Find the stationary distribution.

8.11 Consider a Markov chain with state space $S = \{1, 2, 3, 4\}$ and transition matrix:

$$\mathbf{P} = \begin{bmatrix} 0 & 1 & 0 & 0 \\ 0.5 & 0 & 0.5 & 0 \\ 0 & 0.5 & 0 & 0.5 \\ 0 & 0 & 1 & 0 \end{bmatrix}$$

(a) Draw the state transition diagram.
(b) Determine the period of each state.
(c) Is this chain irreducible? Is it aperiodic?

8.12 A user's session on a website can be in one of four states: Home Page (H), Product Page (P), Cart Page (C), and Checkout Page (K). The transition matrix is:

$$\mathbf{P} = \begin{bmatrix} 0.4 & 0.3 & 0.2 & 0.1 \\ 0.1 & 0.5 & 0.3 & 0.1 \\ 0.1 & 0.2 & 0.6 & 0.1 \\ 0 & 0 & 0 & 1 \end{bmatrix}$$

(a) Draw the state transition diagram.
(b) Identify any absorbing states.
(c) Interpret the meaning of the stationary distribution in this context.

8.13 A simplified model of a hard drive's operation has three states: Active (A), Idle (I), and Failure (F). The transition matrix is:

$$\mathbf{P} = \begin{bmatrix} 0.9 & 0.1 & 0 \\ 0.2 & 0.7 & 0.1 \\ 0 & 0 & 1 \end{bmatrix}$$

(a) Draw the state transition diagram.
(b) What is the probability that the hard drive will eventually fail if it starts in the Active state?
(c) Is this chain ergodic?

8.14 Consider a Markov chain with the following transition matrix:

$$\mathbf{P} = \begin{bmatrix} 0 & 0.6 & 0.4 \\ 0.6 & 0 & 0.4 \\ 0.4 & 0.4 & 0.2 \end{bmatrix}$$

(a) Draw the state transition diagram.
(b) Calculate $\mathbf{P}^2$ and $\mathbf{P}^3$.
(c) Find the stationary distribution.

8.15 A packet in a network can be in one of four states: Transmitting (T), Waiting (W), Collided (C), or Delivered (D). The transition matrix is:

$$\mathbf{P} = \begin{bmatrix} 0.2 & 0.5 & 0.2 & 0.1 \\ 0 & 0.4 & 0.3 & 0.3 \\ 0.6 & 0 & 0.4 & 0 \\ 0 & 0 & 0 & 1 \end{bmatrix}$$

(a) Draw the state transition diagram.
(b) Identify the absorbing state.
(c) What does the stationary distribution signify in this context?

8.16 A simplified weather model has three states: Sunny (S), Cloudy (C), and Rainy (R). The transition matrix is:

$$\mathbf{P} = \begin{bmatrix} 0.7 & 0.2 & 0.1 \\ 0.3 & 0.6 & 0.1 \\ 0.2 & 0.4 & 0.4 \end{bmatrix}$$

(a) Draw the state transition diagram.
(b) If today is Sunny, what is the probability it will be Rainy in 2 days?
(c) Find the stationary distribution.

8.17 Consider a Markov chain with state space $S = \{1, 2, 3, 4\}$ and transition matrix:

$$\mathbf{P} = \begin{bmatrix} 0 & 0.4 & 0.6 & 0 \\ 0.4 & 0 & 0 & 0.6 \\ 0.6 & 0 & 0 & 0.4 \\ 0 & 0.6 & 0.4 & 0 \end{bmatrix}$$

(a) Draw the state transition diagram.
(b) Determine the period of each state.
(c) Is this chain irreducible? Is it aperiodic?

8.18 A customer service system has three states: Available (A), Busy (B), and On Hold (H). The transition matrix is:

$$\mathbf{P} = \begin{bmatrix} 0.4 & 0.4 & 0.2 \\ 0.3 & 0.3 & 0.4 \\ 0.2 & 0.5 & 0.3 \end{bmatrix}$$

(a) Draw the state transition diagram.
(b) Find the stationary distribution.
(c) What does the stationary distribution represent in terms of the system's long-term behavior?

8.19 A simplified model of a file's status in a distributed system has four states: Local (L), Replicating (R), Consistent (C), and Unavailable (U). The transition matrix is:

$$\mathbf{P} = \begin{bmatrix} 0.5 & 0.3 & 0.2 & 0 \\ 0.1 & 0.6 & 0.2 & 0.1 \\ 0 & 0 & 0.8 & 0.2 \\ 0 & 0 & 0 & 1 \end{bmatrix}$$

(a) Draw the state transition diagram.
(b) Identify the absorbing state.
(c) What is the probability that a file will eventually become Unavailable if it starts as Local?

8.20 Consider a Markov chain with state space $S = \{0, 1, 2, 3\}$ and transition matrix:

$$\mathbf{P} = \begin{bmatrix} 0.1 & 0.4 & 0.3 & 0.2 \\ 0.4 & 0.1 & 0.2 & 0.3 \\ 0.3 & 0.2 & 0.1 & 0.4 \\ 0.2 & 0.3 & 0.4 & 0.1 \end{bmatrix}$$

(a) Draw the state transition diagram.
(b) Determine if this chain is irreducible and aperiodic.
(c) Find the stationary distribution.

Chapter 9
Poisson Processes

The Poisson process is a fundamental concept in probability theory and stochastic processes, widely used to model the occurrence of random events over time. While the classical Poisson process operates in continuous time, many real-world phenomena are observed or recorded in discrete time intervals. This chapter focuses on the *discrete-time Poisson process*, an adaptation of the Poisson process for discrete-time settings. We will explore its definition, properties, and applications, particularly with illustrative numerical examples.

9.1 Basic Definitions and Related Equations

Discrete-Time Setting

In a discrete-time setting, time is divided into equal-length intervals (e.g., seconds, minutes, days). We observe the number of events occurring in each interval. Let N_k represent the number of events that have occurred up to time step k, where $k = 0, 1, 2, ...$

Discrete-Time Poisson Process Definition

A discrete-time stochastic process $\{N_k, k = 0, 1, 2, ...\}$ is called a discrete-time Poisson process with rate λ if it satisfies the following conditions:

(i) $N_0 = 0$, i.e. the process starts at 0.
(ii) The number of events occurring in each interval is independent of the number of events in any other non-overlapping interval.

M. Catak and T. Allahviranloo, *Discrete Random Processes*, Studies in Systems, Decision and Control 644, https://doi.org/10.1007/978-3-032-08630-3_9

(iii) The probability of n events occurring in any interval of unit length is given by:

$$P(N_k - N_{k-1} = n) = \frac{\lambda^n e^{-\lambda}}{n!}, \quad n = 0, 1, 2, ...$$

This means that the increments $N_k - N_{k-1}$ are independent and identically distributed (i.i.d.) random variables, each following a Poisson distribution with parameter λ.

Poisson Distribution

Recall that a random variable X follows a Poisson distribution with parameter $\lambda > 0$ if its probability mass function (PMF) is:

$$P(X = k) = \frac{\lambda^k e^{-\lambda}}{k!}, \quad k = 0, 1, 2, ... \tag{9.1}$$

The mean and variance of a Poisson random variable are both equal to λ:

$$E[X] = \lambda$$

$$Var[X] = \lambda$$

Hence, the probability of observing exactly n events up to time step k is given by the Poisson distribution:

$$P(N_k = n) = \frac{(\lambda k)^n e^{-\lambda k}}{n!}, \quad n = 0, 1, 2, ... \tag{9.2}$$

9.2 Properties of a Poisson Process

Independent Increments

The independent increments property is a fundamental defining characteristic of the Poisson process. It states that the number of events occurring in any one time interval is statistically independent of the number of events occurring in any other non-overlapping time interval.

Let $0 \leq k_1 < k_2 < k_3 < k_4$. Then, the random variables $N_{k_2} - N_{k_1}$, number of events in the interval $(k_1, k_2])$, and $N_{k_4} - N_{k_3}$, number of events in the interval $(k_3, k_4])$ are independent. This property is not derived but rather assumed as part of the definition of the discrete-time Poisson process. The definition explicitly states that the number of events occurring in each interval is independent of the number of events in any other non-overlapping interval. Therefore, it's a foundational axiom of the model.

Stationary Increments

Stationary increments mean that the probability distribution of the number of events within an interval depends only on the length of that interval, not on where the interval starts. This implies that the process is homogeneous in time.

For any integers k and h (where $h > 0$), the random variable $N_{k+h} - N_k$ has the same distribution as $N_h - N_0$. Since $N_0 = 0$, this is the same as saying $N_{k+h} - N_k$ has the same distribution as N_h. This property stems from the definition of the process. The probability of n events occurring in any interval of unit length is given by the Poisson distribution with parameter λ. Therefore, for an interval of length h, the number of events follows a Poisson distribution with parameter λh, regardless of the interval's starting point.

$$P(N_{k+h} - N_k = n) = \frac{(\lambda h)^n e^{-\lambda h}}{n!} = P(N_h = n) \tag{9.3}$$

Mean and Variance

We can derive the mean and variance of the discrete-time Poisson process from the properties of the Poisson distribution.

It is known that N_1 follows a Poisson distribution with parameter λ. Therefore, $E[N_1] = \lambda$. Since the increments are stationary, N_k can be seen as the sum of k independent Poisson random variables, each with mean λ:

$$N_k = (N_1 - N_0) + (N_2 - N_1) + ... + (N_k - N_{k-1})$$

Using the linearity of expectation operator:

$$E[N_k] = E[N_1 - N_0] + E[N_2 - N_1] + ... + E[N_k - N_{k-1}] = \lambda + \lambda + ... + \lambda = k\lambda \tag{9.4}$$

Similarly, $Var[N_1] = \lambda$. Since the increments are independent, the variance of the sum is the sum of the variances:

$$Var[N_k] = Var[N_1 - N_0] + Var[N_2 - N_1] + ... + Var[N_k - N_{k-1}] = \lambda + \lambda + ... + \lambda = k\lambda \tag{9.5}$$

Superposition

When one combine multiple independent Poisson processes, the resulting process is also Poisson, with its rate being the sum of the individual rates.

Let $N_k^{(1)}, N_k^{(2)}, ..., N_k^{(m)}$ be m independent discrete-time Poisson processes with rates $\lambda_1, \lambda_2, ..., \lambda_m$, respectively. Let $N_k = N_k^{(1)} + N_k^{(2)} + ... + N_k^{(m)}$. It should be shown that N_k is also a discrete-time Poisson process with rate $\lambda = \lambda_1 + \lambda_2 + ... + \lambda_m$.

The sum of independent Poisson random variables is also a Poisson random variable. Therefore, N_k follows a Poisson distribution. The rate parameter of the sum is the sum of the individual rates.

$$P(N_k = n) = \frac{(\lambda_1 k + \lambda_2 k + ... + \lambda_m k)^n e^{-(\lambda_1 k + \lambda_2 k + ... + \lambda_m k)}}{n!} = \frac{(\lambda k)^n e^{-\lambda k}}{n!} \tag{9.6}$$

where $\lambda = \lambda_1 + \lambda_2 + ... + \lambda_m$.

The independent increments property of individual processes implies the independent increments property of the combined process.

Decomposition

If we *thin* a Poisson process by classifying events into types, each type's counting process is also Poisson. *Thinning* is a way of creating new Poisson processes from existing ones by randomly selecting which events to keep. It is a useful concept for modeling situations where you have a random stream of events, but you are only interested in a specific subset of those events. For instance, a stream of emails arriving in your inbox, modeled as a Poisson process. Then, you decide to categorize each email such as *Important* or *Spam*. Each email is independently labeled as *Important* with probability p and *Spam* with probability $1 - p$. The process of counting only the *Important* emails is a *thinned* Poisson process, since a subset of the original email events will only be counted.

Let N_k be a discrete-time Poisson process with rate λ. Suppose each event is classified as type i with probability p_i, where $i = 1, 2, ..., m$ and $\sum_{i=1}^{m} p_i = 1$. Let $N_k^{(i)}$ be the process counting events of type i.

The probability of having n events of type i in an interval of length L follows a binomial distribution within a Poisson distribution framework. It can be shown with some probability manipulations that as the rate increases and the probability decreases appropriately, this converges to a Poisson distribution with rate λp_i.

$$P(N_k^{(i)} = n) = \frac{(\lambda k p_i)^n e^{-(\lambda k p_i)}}{n!} \tag{9.7}$$

The independent increments property of the original process ensures the independence of increments for the decomposed processes.

9.3 Applications of Poisson Processes

9.3.1 *Modeling Website Traffic*

Consider a website that receives user requests. It can be modelled that the number of requests received per minute as a discrete-time Poisson process. Suppose the website receives an average of 120 user requests per minute. This process can be realized by means of a discrete-time Poisson process where each time interval is 1 minute and the rate $\lambda = 120$ requests/minute. Then, for example the probability of receiving exactly 100 requests in a given minute is calculated:

$$P(N_1 = 100) = \frac{120^{100}e^{-120}}{100!} \approx 0.0276$$

The probability of receiving more than 150 requests in a minute is:

$$P(N_1 > 150) = 1 - \sum_{n=0}^{150} \frac{120^n e^{-120}}{n!} \approx 0.1085$$

The expected number of requests in 5 minutes is derived:

$$E[N_5] = \lambda \times 5 = 120 \times 5 = 600 \text{ requests}$$

9.3.2 *Modeling Particle Counts in an Experiment*

In a physics experiment, the number of particles detected by a counter in a fixed time interval can often be modeled as a Poisson process. For instance, an appropriate counter detects radioactive particles. Assume that the average count rate is 5 particles per second. One can model this as a discrete-time Poisson process with a 1 second interval and $\lambda = 5$ particles/second.

What is the probability of detecting exactly 8 particles in a second?

$$P(N_1 = 8) = \frac{5^8 e^{-5}}{8!} \approx 0.0653$$

What is the probability of detecting between 3 and 7 particles (inclusive) in a second?

$$P(3 \le N_1 \le 7) = \sum_{n=3}^{7} \frac{5^n e^{-5}}{n!} \approx 0.6160$$

If we extend the counting interval to 10 seconds, what is the variance of the number of particles·detected?

$$Var[N_{10}] = \lambda \times 10 = 5 \times 10 = 50$$

9.3.3 Modeling Phone Calls in a Call Center

The number of incoming calls to a call center per unit of time can be modeled as a discrete-time Poisson process.

Let assume that a call center receives an average of 3 calls per minute. We model this with $\lambda = 3$ calls/minute.

What is the probability that the call center receives no calls in a minute?

$$P(N_1 = 0) = \frac{3^0 e^{-3}}{0!} = e^{-3} \approx 0.0498$$

What is the probability that the call center receives exactly 5 calls in 2 minutes? Since the rate for 2 minutes would be $\lambda \times 2 = 6$,

$$P(N_2 = 5) = \frac{6^5 e^{-6}}{5!} \approx 0.1606$$

If the call center receives calls from two independent sources, one averaging 2 calls/minute and the other averaging 1 call/minute, what is the combined rate? The combined rate would be $\lambda = 2 + 1 = 3$ calls/minute, illustrating the superposition property.

Problems

9.1 Events occur according to a discrete-time Poisson process with a rate of $\lambda = 4$ events per hour.

(a) What is the probability of exactly 5 events occurring in a given hour?
(b) What is the probability of at least 2 events occurring in a given hour?
(c) What is the probability of no events occurring in a 30-minute period?

9.2 The number of customers entering a store each day follows a discrete-time Poisson process with a rate of $\lambda = 20$ customers per day.

(a) What is the expected number of customers entering the store in a 5-day period?

(b) What is the variance of the number of customers entering the store in a 10-day period?
(c) What is the standard deviation of the number of customers entering the store in a half-day (12-hour) period? (Assume the rate scales proportionally.)

9.3 Two independent discrete-time Poisson processes count the number of emails received by a user. Process 1 has a rate of $\lambda_1 = 10$ emails per hour, and Process 2 has a rate of $\lambda_2 = 15$ emails per hour.

(a) What is the rate of the combined process that counts all emails received?
(b) What is the probability that the user receives a total of exactly 20 emails in a given hour?
(c) What is the expected number of total emails received in 3 hours?

9.4 A manufacturing process produces items, and defects occur according to a discrete-time Poisson process with a rate of $\lambda = 0.5$ defects per item. Each defect is independently classified as either type A or type B, with probabilities $p_A = 0.4$ and $p_B = 0.6$.

(a) What is the rate of the Poisson process counting type A defects?
(b) What is the probability that an item has exactly 1 type B defect?
(c) What is the expected number of type A defects in 10 items?

9.5 Events occur according to a discrete-time Poisson process with a rate of $\lambda = 2$ events per minute. Let T be the time (in minutes) until the first event occurs.

(a) What is the probability that the first event occurs in the first minute (i.e., $P(T = 1)$)?
(b) What is the probability that the first event occurs after 2 minutes (i.e., $P(T > 2)$)?
(c) What is the probability that the first event occurs at the 3rd minute (i.e., $P(T = 3)$)?

9.6 Customers arrive at a service counter according to a discrete-time Poisson process with a rate of $\lambda = 5$ customers per hour.

(a) What is the probability that exactly 3 customers arrive in an hour given that at least 1 customer arrives in that hour?
(b) What is the probability that exactly 2 customers arrive in the first half-hour given that exactly 4 customers arrive in the whole hour?

9.7 The number of calls arriving at a telephone exchange in a short time interval can be approximated by a discrete-time Poisson process. Suppose the rate is $\lambda = 0.1$ calls per second.

(a) Use the Poisson approximation to calculate the probability that exactly 2 calls arrive in 10 seconds.

(b) Compare your answer in (a) to the exact probability calculated using the binomial distribution. (Hint: Consider each second as a trial.)
(c) Explain why the Poisson distribution is a suitable approximation in this scenario.

9.8 A compound Poisson process is a process where the number of events follows a Poisson process, and each event has an associated random variable. For example, customers arrive at a store according to a discrete-time Poisson process, and each customer spends a random amount of money.

Suppose customers arrive at a store according to a discrete-time Poisson process with a rate of $\lambda = 4$ customers per hour. The amount of money spent by each customer is independent and follows a distribution with a mean of \$20 and a variance of \$100. Let S_t be the total amount of money spent by customers in t hours.

(a) What is the expected total amount of money spent by customers in 2 hours?
(b) What is the variance of the total amount of money spent by customers in 2 hours?

9.9 In a non-homogeneous Poisson process, the rate λ is a function of time, $\lambda(t)$. Explain how the concept of a discrete-time Poisson process can be extended to approximate a discrete-time version of a non-homogeneous Poisson process. Describe how you would adapt the calculations for probabilities and expected values.

9.10 Describe how you would simulate a discrete-time Poisson process with a given rate λ using a computer software. Provide pseudocode or a brief description of the algorithm.

9.11 Messages arrive at a communication server according to a discrete-time Poisson process with a rate of $\lambda = 7$ messages per minute.

(a) What is the probability that exactly 10 messages arrive in a 2-minute interval?
(b) What is the probability that at least one message arrives in a 10-second interval? (Assume the rate scales proportionally.)
(c) If it is known that at least one message arrives in a minute, what is the probability that exactly 3 messages arrive in that minute?

9.12 Errors occur in a data transmission system according to a discrete-time Poisson process. The average number of errors is 2 per 1000 bits transmitted.

(a) What is the probability of no errors in the transmission of 100 bits?
(b) What is the probability of more than 5 errors in the transmission of 2000 bits?
(c) If we transmit N bits, what is the approximate probability of no errors, and how large should N be for this probability to be at least 0.99?

9.13 Let X_t be a discrete-time Poisson process with rate λ.

(a) Find the probability mass function of X_t.
(b) Compute the mean and variance of X_t.
(c) Show that the increments of the Poisson process are independent.

9.14 Consider two independent discrete-time Poisson processes, X_t and Y_t, with rates λ_1 and λ_2, respectively.

(a) What is the distribution of $X_t + Y_t$?
(b) Find the probability $P(X_t = k|X_t + Y_t = n)$ for $0 \leq k \leq n$.
(c) Explain the significance of the result in (b).

9.15 A website receives hits according to a discrete-time Poisson process with an average of 10 hits per minute. The server can handle a maximum of 15 hits per minute.

(a) What is the probability that the server is overloaded in any given minute?
(b) What is the probability that the server is overloaded in at least 2 out of 5 consecutive minutes?
(c) If the server capacity is increased to 20 hits per minute, what is the new probability of overload?

9.16 Accidents occur at an intersection according to a discrete-time Poisson process with a rate of 1 accident per week.

(a) What is the probability of exactly 4 accidents occurring in a month (4 weeks)?
(b) What is the probability of no accidents occurring in a quarter (13 weeks)?
(c) What is the probability that the 5th accident occurs in the 10th week?

9.17 A store sells a particular product. The demand for the product follows a discrete-time Poisson process with a rate of λ units per day. The store orders S units of the product every day.

(a) What is the probability that the demand exceeds the supply on any given day?
(b) If $\lambda = 5$ and $S = 7$, what is the probability that the demand exceeds the supply?
(c) How large should S be so that the probability of demand exceeding supply is less than 0.01?

9.18 Consider a discrete-time Poisson process with rate λ. Let T_k be the time (in discrete units) until the k-th event occurs.

(a) What is the distribution of T_1?
(b) What is the expected value of T_k?
(c) What is the variance of T_k?

9.19 In a certain factory, machines fail according to a discrete-time Poisson process with a rate of 0.2 failures per day. The factory has 10 machines.

(a) What is the probability that exactly 3 machines fail in a week (7 days)?
(b) What is the probability that at least one machine fails in a day?
(c) If a machine fails, it is repaired immediately. What is the expected number of repairs in a month (30 days)?

9.20 Describe how you would test if a given set of discrete-time event data can be reasonably modeled as a discrete-time Poisson process. What statistical tests or methods could you use? Discuss the assumptions that need to be checked.

Chapter 10
Ornstein-Uhlenbeck Process

The Ornstein-Uhlenbeck (OU) process is a cornerstone in the study of stochastic processes, particularly in modeling phenomena that exhibit a tendency to revert towards a long-term mean. Originally developed to describe the velocity of a Brownian particle under the influence of friction, its applications have broadened significantly, spanning fields such as finance (modeling interest rates and commodity prices), biology (describing population dynamics), and signal processing (noise modeling).

While the continuous-time OU process is often the initial focus, many real-world observations and computational simulations naturally occur in discrete time steps. This chapter delves into the discrete-time counterpart of the OU process, exploring its definition, key properties, and illustrating its behavior through examples and realizations. Understanding the discrete-time OU process provides a powerful tool for analyzing time series data that exhibit mean-reverting characteristics in a sampled or inherently discrete setting.

10.1 General Definitions and Related Mathematical Equations

The continuous-time Ornstein-Uhlenbeck process, $X(t)$, is typically defined by the following stochastic differential equation (SDE):

$$dX(t) = \theta(\mu - X(t))dt + \sigma dW(t) \tag{10.1}$$

M. Catak and T. Allahviranloo, *Discrete Random Processes*, Studies in Systems, Decision and Control 644, https://doi.org/10.1007/978-3-032-08630-3_10

where:

$\theta > 0$ is the rate of mean reversion, indicating the strength of the pull towards the mean.
μ is the long-term mean or equilibrium level.
$\sigma > 0$ is the volatility parameter, representing the magnitude of the random fluctuations.
$W(t)$ is a standard Wiener process (Brownian motion).

To transition to the discrete-time setting, we consider the process at discrete time points $t_0, t_1, t_2, \ldots$, where the time interval between consecutive points is constant, $\Delta t = t_{i+1} - t_i$. Let $X_n = X(t_n)$ denote the value of the process at time $t_n = n\Delta t$.

One way to derive the discrete-time OU process is by considering the exact solution of the continuous-time SDE over a small time interval Δt. The solution to the continuous-time OU SDE is given by:

$$X(t) = X(s)e^{-\theta(t-s)} + \mu(1 - e^{-\theta(t-s)}) + \sigma \int_s^t e^{-\theta(t-u)} dW(u) \tag{10.2}$$

Setting $t = (n+1)\Delta t$ and $s = n\Delta t$, we obtain the discrete-time evolution:

$$X_{n+1} = X_n e^{-\theta\Delta t} + \mu(1 - e^{-\theta\Delta t}) + \sigma \int_{n\Delta t}^{(n+1)\Delta t} e^{-\theta((n+1)\Delta t - u)} dW(u) \tag{10.3}$$

Let $\alpha = e^{-\theta\Delta t}$ and $\beta = \mu(1 - e^{-\theta\Delta t}) = \mu(1 - \alpha)$. Then the equation becomes:

$$X_{n+1} = \alpha X_n + \beta + \sigma \int_{n\Delta t}^{(n+1)\Delta t} e^{-\theta((n+1)\Delta t - u)} dW(u) \tag{10.4}$$

The integral term represents a normally distributed random variable with mean zero. To find its variance:

$$Var\left(\int_{n\Delta t}^{(n+1)\Delta t} e^{-\theta((n+1)\Delta t - u)} dW(u)\right)$$

$$= \int_{n\Delta t}^{(n+1)\Delta t} \left(e^{-\theta((n+1)\Delta t - u)}\right)^2 du$$

Let $v = (n+1)\Delta t - u$, then $dv = -du$. When $u = n\Delta t$, $v = \Delta t$, and when $u = (n+1)\Delta t$, $v = 0$. Thus, the integral becomes:

$$\int_0^{\Delta t} e^{-2\theta v} dv = \left[-\frac{1}{2\theta} e^{-2\theta v} \right]_0^{\Delta t}$$

$$= \frac{1}{2\theta}(1 - e^{-2\theta \Delta t})$$

$$= \frac{1 - \alpha^2}{2\theta}$$

Therefore, the variance of the random term is $\sigma^2 \frac{1-\alpha^2}{2\theta}$.

Let $\sigma_d^2 = \sigma^2 \frac{1-\alpha^2}{2\theta}$. One can then define the discrete-time Ornstein-Uhlenbeck process as:

$$X_{n+1} = \alpha X_n + \beta + \sigma_d \epsilon_{n+1} \tag{10.5}$$

where $\epsilon_{n+1} \sim \mathcal{N}(0, 1)$ is a sequence of independent and identically distributed (i.i.d.) standard normal random variables.

The discrete-time Ornstein-Uhlenbeck process is characterized by three crucial parameters that directly relate to the underlying continuous-time process. The first is the persistence parameter, $\alpha = e^{-\theta \Delta t}$. Given that the mean reversion rate θ and the time step Δt are both positive, α will always lie between 0 and 1 ($0 < \alpha < 1$). The magnitude of α dictates the memory of the process; when α is close to 1, the process exhibits strong persistence, meaning its current value is heavily influenced by the previous one. Conversely, an α closer to 0 indicates weaker persistence.

The parameter β, defined as $\mu(1 - e^{-\theta \Delta t})$ or equivalently $\mu(1 - \alpha)$, is directly linked to the long-term mean μ of the process. Finally, the standard deviation of the noise term in the discrete-time process, $\sigma_d = \sigma \sqrt{\frac{1-e^{-2\theta \Delta t}}{2\theta}}$, which can also be expressed as $\sigma \sqrt{\frac{1-\alpha^2}{2\theta}}$, governs the magnitude of the random fluctuations at each time step and is derived from the volatility σ of the continuous process.

Stationary Distribution

If the process is started with an initial condition X_0, its distribution will evolve over time. However, if $|\alpha| < 1$,which is always the case here since $0 < \alpha < 1$, the discrete-time OU process has a stationary distribution. As $n \to \infty$, the distribution of X_n converges to a normal distribution with:

Mean: $E[X_\infty] = \frac{\beta}{1-\alpha} = \frac{\mu(1-\alpha)}{1-\alpha} = \mu$

Variance: $Var(X_\infty) = \frac{\sigma_d^2}{1-\alpha^2} = \frac{\sigma^2(1-\alpha^2)/(2\theta)}{1-\alpha^2} = \frac{\sigma^2}{2\theta}$

Notice that the stationary mean matches the long-term mean μ of the continuous-time process. The stationary variance is also related to the parameters of the continuous-time process.

Examples with Realizations

Let's explore the behavior of the discrete-time OU process with some specific parameter values and visualize a few realizations.

Strong Mean Reversion
Consider an OU Process having the following parameters:

$\mu = 0$ (long-term mean is zero)
$\theta = 2$ (strong mean reversion)
$\sigma = 1$ (moderate volatility)
$\Delta t = 0.1$ (time step)

One can calculate the discrete-time parameters:

$\alpha = e^{-2\times 0.1} = e^{-0.2} \approx 0.8187$

$\beta = 0 \times (1 - 0.8187) = 0$

$\sigma_d = 1 \times \sqrt{\frac{1-(0.8187)^2}{2\times 2}} = \sqrt{\frac{1-0.6703}{4}} = \sqrt{\frac{0.3297}{4}} \approx \sqrt{0.0824} \approx 0.2871$

The discrete-time process is then:

$$X_{n+1} = 0.8187X_n + 0.2871\epsilon_{n+1}$$

Let's simulate a realization starting from $X_0 = 1$ as displayed in Table 10.1. It would be observed from Table 10.1, the process tends to fluctuate around the mean of 0, with the strong mean reversion pulling it back towards zero after deviations.

Table 10.1 Realization of a discrete-time OU process with strong mean reversion

n	X_n	ϵ_{n+1}	$X_{n+1} = 0.8187X_n + 0.2871\epsilon_{n+1}$
0	1.0000	0.50	0.9623
1	0.9623	–0.20	0.7305
2	0.7305	1.10	0.9138
3	0.9138	–0.80	0.5184
4	0.5184	0.35	0.5249
5	0.5249	–1.50	–0.0009
6	–0.0009	0.75	0.2146
7	0.2146	–0.55	0.0178
8	0.0178	0.90	0.2730
9	0.2730	–0.15	0.1804
10	0.1804	–0.65	–0.0389

Weak Mean Reversion

Now, let's consider a case with weaker mean reversion:

$\mu = 5$
$\theta = 0.1$ (weak mean reversion)
$\sigma = 0.5$
$\Delta t = 0.1$

The discrete-time parameters are:

$\alpha = e^{-0.1\times 0.1} = e^{-0.01} \approx 0.9900$

$\beta = 5 \times (1 - 0.9900) = 5 \times 0.01 = 0.05$

$\sigma_d = 0.5 \times \sqrt{\frac{1-(0.9900)^2}{2\times 0.1}} = 0.5 \times \sqrt{\frac{1-0.9801}{0.2}} = 0.5 \times \sqrt{\frac{0.0199}{0.2}} \approx 0.5 \times \sqrt{0.0995} \approx 0.5 \times 0.3154 \approx 0.1577$

The discrete-time process is:

$$X_{n+1} = 0.9900X_n + 0.05 + 0.1577\epsilon_{n+1}$$

Starting from $X_0 = 3$, a possible realization results are given in Table 10.2.
In this case, with α close to 1, the process exhibits stronger persistence, and the fluctuations around the mean of 5 are more gradual as depicted in Table 10.2.

Table 10.2 Realization of a discrete-time OU process with weak mean reversion

n	X_n	ϵ_{n+1}	$X_{n+1} = 0.9900X_n + 0.05 + 0.1577\epsilon_{n+1}$
0	3.0000	1.20	3.2392
1	3.2392	–0.50	3.2603
2	3.2603	0.80	3.5585
3	3.5585	–0.10	3.5521
4	3.5521	0.65	3.6780
5	3.6780	–0.25	3.6888
6	3.6888	0.40	3.7052
7	3.7052	0.15	3.7180
8	3.7180	–0.70	3.6293
9	3.6293	0.95	3.8216
10	3.8216	–0.05	3.8122
11	3.8122	0.30	3.8561
12	3.8561	–0.45	3.7913
13	3.7913	0.60	3.8195
14	3.8195	–0.10	3.8118
15	3.8118	0.20	3.8506
16	3.8506	0.05	3.8530
17	3.8530	–0.35	3.7981
18	3.7981	0.70	3.8103
19	3.8103	–0.20	3.7669
20	3.7669	0.55	3.8038

10.2 Application of the Ornstein-Uhlenbeck Process in Financial Mathematics: Modeling Short-Term Interest Rates

10.2.1 Mean Reversion in Short-Term Interest Rates

Short-term interest rates, unlike asset prices that often exhibit trending behavior, tend to fluctuate around a long-term equilibrium level. This phenomenon, known as mean reversion, is driven by central bank policies, macroeconomic conditions, and market forces that prevent rates from deviating indefinitely from their perceived "normal" level. For instance, if short-term rates rise significantly above the central bank's target, monetary policy tightening might ease, and market participants might anticipate a future decrease, pushing rates back down. Conversely, very low rates might spur inflation concerns and expectations of future rate hikes. Modeling this mean-reverting behavior is crucial for pricing interest rate-sensitive financial instruments and managing interest rate risk.

10.2.2 The Vasicek Model: Mathematical Formulation

The Vasicek model employs the Ornstein-Uhlenbeck (OU) process to describe the evolution of the instantaneous short-term interest rate, $r(t)$, through the following stochastic differential equation (SDE) [1]:

$$dr(t) = a(b - r(t))dt + \sigma dW(t)$$

where:

$r(t)$: Short-term interest rate at time t.
$a > 0$: Speed of mean reversion (rate of adjustment towards b).
$b > 0$: Long-term mean level of the interest rate (equilibrium rate).
$\sigma > 0$: Volatility of the interest rate process (magnitude of random fluctuations).
$dW(t)$: Increment of a standard Wiener process (Brownian motion, representing random market shocks).

The discrete-time representation, useful for simulation and empirical analysis with discrete data, is given by:

$$r_{n+1} = r_n e^{-a\Delta t} + b(1 - e^{-a\Delta t}) + \sigma\sqrt{\frac{1 - e^{-2a\Delta t}}{2a}}\epsilon_{n+1}$$

where Δt is the time step and $\epsilon_{n+1} \sim \mathcal{N}(0, 1)$ is a standard normal random variable.

Illustrative Results: Simulation of the Federal Funds Rate

Consider the U.S. Federal Funds Rate (FFR) as an example. Assume the following calibrated parameters: long-term mean $b = 0.025$ (2.5%), speed of mean reversion $a = 0.3$, and volatility $\sigma = 0.01$. Using a daily time step ($\Delta t = 1/252$) and starting from $r_0 = 0.02$ (2%), we can simulate the path of the FFR (Table 10.3).

The simulated FFR fluctuates stochastically but exhibits a tendency to revert towards the long-term mean of 2.5%. The speed of this reversion and the magnitude of the fluctuations are governed by the parameters a and σ, respectively. Over time, the simulated rate oscillates around the mean, reflecting the balance between the pull towards equilibrium and random market shocks.

Discussion of Model Efficiency and Limitations

The Vasicek model, rooted in the Ornstein-Uhlenbeck process, presents several notable benefits for modeling the dynamics of short-term interest rates. Firstly, it inherently captures the crucial characteristic of mean reversion, an empirically

Table 10.3 Simulated path of the federal funds rate (illustrative)

Day (n)	Simulated FFR (r_n)
0	0.02000
50	0.02115
100	0.02302
150	0.02418
200	0.02535
250	0.02489
300	0.02512
350	0.02601
400	0.02455
450	0.02578
500	0.02495

observed phenomenon where interest rates tend to gravitate back towards a long-term average. This feature distinguishes it favorably from models based on geometric Brownian motion, which do not incorporate such a stabilizing force. Secondly, the model's analytical tractability is a significant advantage. It allows for the derivation of closed-form solutions for the prices of zero-coupon bonds and certain interest rate derivatives, thereby simplifying the processes of valuation and risk management for financial practitioners. Finally, the parameters of the Vasicek model, the speed of mean reversion (a), the long-term mean (b), and the volatility (σ), possess clear and intuitive economic interpretations. This transparency aids in the model's calibration to market data and facilitates a deeper understanding of its implications for interest rate behavior.

However, the Vasicek model is not without its limitations. One notable drawback is the theoretical possibility of negative interest rates arising from the Ornstein-Uhlenbeck process. While historically viewed as unrealistic, the occurrence of negative rates in some economies has somewhat altered this perspective, though the basic Vasicek framework doesn't inherently prevent such outcomes and might necessitate adjustments in specific contexts. Another significant limitation lies in the assumption of constant volatility (σ). In reality, the volatility of interest rates is often observed to be time-varying, influenced by prevailing market conditions and economic uncertainty. More advanced models incorporating stochastic volatility have been developed to address this deficiency. Furthermore, the model's representation of mean reversion as a linear function of the deviation from the mean might be a simplification of more complex, potentially non-linear dynamics present in actual interest rate movements.

Despite these acknowledged limitations, the Vasicek model remains a foundational and extensively utilized framework for comprehending and modeling the behavior of short-term interest rates. Its relative simplicity and the analytical solutions it provides render it a valuable tool within financial mathematics. It often serves as an initial stepping stone for understanding more intricate models designed to overcome its shortcomings and capture the nuances of interest rate dynamics with greater precision.

To conclude, the discrete-time Ornstein-Uhlenbeck process provides a flexible and widely applicable framework for modeling mean-reverting time series data. By varying the parameters θ and σ, and the time step Δt, one can observe how these factors influence the characteristics of the discrete-time OU process, such as the rate of decay of autocorrelation, the variance of the fluctuations, and the speed at which the process reverts to its long-term mean.

Its parameters are directly related to the parameters of its continuous-time counterpart, and its behavior can be effectively analyzed and simulated using the derived equations. Understanding its properties and observing its realizations offers valuable insights into the dynamics of systems exhibiting a tendency towards equilibrium under the influence of random perturbations.

Problems

10.1 Consider a discrete-time Ornstein-Uhlenbeck process given by $X_{n+1} = \alpha X_n + \beta + \sigma_d \epsilon_{n+1}$, where $\alpha = 0.8$, $\beta = 1$, and $\sigma_d = 0.3$. Assume $\epsilon_n \sim \mathcal{N}(0, 1)$ are i.i.d.

(a) If $X_0 = 0$, what is the expected value of X_1?
(b) If $X_0 = 0$, what is the variance of X_1?
(c) If $X_1 = 2$, what is the expected value of X_2?

10.2 A discrete-time OU process is defined by $X_{n+1} = 0.9X_n + 0.5 + 0.2\epsilon_{n+1}$.

(a) What is the long-term mean of this process?
(b) What is the variance of the stationary distribution of this process?
(c) If the process is currently at $X_n = 3$, what is the probability distribution of X_{n+1}? Specify the mean and variance.

10.3 The parameters of a continuous-time OU process are $\theta = 0.5$, $\mu = 10$, and $\sigma = 2$. The process is observed at discrete time intervals of $\Delta t = 0.2$.

(a) Calculate the parameters α and β for the corresponding discrete-time OU process.
(b) Calculate the parameter σ_d for the corresponding discrete-time OU process.

10.4 Consider two discrete-time OU processes, Process-1 with $\alpha_1 = 0.7$ and Process-2 with $\alpha_2 = 0.95$ (all other parameters being the same).

(a) Which process exhibits stronger persistence? Explain your reasoning.
(b) Which process would you expect to return to its long-term mean faster after a large deviation? Explain your reasoning.

10.5 A short-term interest rate is modeled by a discrete-time Vasicek model: $r_{n+1} = \alpha r_n + \beta + \sigma_d \epsilon_{n+1}$. Suppose $\alpha = 0.98$, $\beta = 0.005$, and $\sigma_d = 0.001$.

(a) What is the implied long-term mean interest rate?

(b) If the current interest rate is $r_n = 0.03$, what is the expected value of the interest rate in the next period?

10.6 Consider a discrete-time OU process $X_{n+1} = \alpha X_n + \beta + \sigma_d \epsilon_{n+1}$. If the variance of the stationary distribution is $\frac{\sigma^2}{2\theta}$, and we know $\alpha = e^{-\theta \Delta t}$, $\sigma_d^2 = \sigma^2 \frac{1-\alpha^2}{2\theta}$, derive the relationship between the parameters of the discrete-time process (σ_d and α) and the variance of the stationary distribution.

10.7 Simulate a single step of a discrete-time OU process with parameters $\alpha = 0.6$, $\beta = -1$, and $\sigma_d = 0.4$, starting from $X_0 = 2$. Assume a random draw from the standard normal distribution gives $\epsilon_1 = 1.5$. What is the value of X_1?

10.8 In the context of modeling commodity prices with a mean-reverting process, what economic factors might justify the use of an OU process rather than a geometric Brownian motion? How would the parameters of the OU process relate to these economic factors?

10.9 Consider a discrete-time OU process. If the time step Δt is increased while the underlying continuous-time parameters θ, μ, and σ remain constant, how would you expect the parameters α, β, and σ_d of the discrete-time process to change?

10.10 Explain the difference in the behavior of realizations of a discrete-time OU process with strong mean reversion (large θ, α close to 0 for a fixed Δt) versus weak mean reversion (small θ, α close to 1 for a fixed Δt), assuming they have the same long-term mean and volatility.

10.11 A discrete-time OU process is given by $X_{n+1} = 0.7X_n - 0.2 + 0.5\epsilon_{n+1}$.

(a) What is the expected change in the process from time n to $n + 1$, given $X_n = x$?
(b) What is the probability distribution of the noise term affecting the process at each step?
(c) If the process starts at $X_0 = -1$, what is the expected value of X_2?

10.12 Consider a discrete-time OU process $X_{n+1} = \alpha X_n + \beta + \sigma_d \epsilon_{n+1}$. If the long-term mean μ is known, express β in terms of μ and α.

10.13 The continuous-time OU process for the velocity of a particle is given by $dV(t) = -\gamma V(t)dt + \eta dW(t)$, where $\gamma > 0$ is the friction coefficient and $\eta > 0$ is the noise strength. This is a special case with $\mu = 0$ and $\theta = \gamma$, $\sigma = \eta$. Derive the parameters α and σ_d for the discrete-time version of this velocity process with a time step Δt.

10.14 Explain intuitively why a higher value of the mean reversion rate θ in the continuous-time OU process leads to a smaller value of the persistence parameter α in the discrete-time version (for a fixed Δt).

10.15 A discrete-time OU process has a stationary distribution with a mean of 5 and a variance of 2. If the persistence parameter is $\alpha = 0.8$, what is the value of β?

10.16 Consider the discrete-time OU process $X_{n+1} = \alpha X_n + \beta + \sigma_d \epsilon_{n+1}$. If you have a time series data that you suspect follows this process, how could you try to estimate the parameters α, β, and σ_d from the data? (A qualitative answer is sufficient).

10.17 Simulate two steps of a discrete-time OU process with $\alpha = 0.95$, $\beta = 1.5$, and $\sigma_d = 0.2$, starting from $X_0 = 0$. Assume the random draws from the standard normal distribution are $\epsilon_1 = -0.8$ and $\epsilon_2 = 0.5$. What are the values of X_1 and X_2?

10.18 In the context of modeling population size with an OU-like process (where the population tends towards a carrying capacity), how would the parameters relate to the biological factors influencing the population dynamics?

10.19 Consider a continuous-time OU process with parameters θ, μ, and σ. If the time step for discretization Δt becomes very small ($\Delta t \to 0$), what do the parameters α, β, and σ_d of the discrete-time process approach? (Hint: Use Taylor series expansions for e^x around $x = 0$).

10.20 Compare and contrast the long-term behavior (as $n \to \infty$) of a discrete-time OU process with $|\alpha| < 1$ and a process with $|\alpha| = 1$ (a random walk with drift). What is the key difference regarding their stationary properties?

Reference

1. Svoboda, Simona. 2004. *The vasicek model, 3–17*. Interest Rate Modelling: Springer.

Solutions

Problems of Chapter 1

1.1 (a) Non-linear
(b) Linear
(c) Linear
(d) Non-linear

1.2 (a) Causal
(b) Non-causal
(c) Causal
(d) Causal

1.3 (a) Time-invariant
(b) Time-varying
(c) Time-invariant
(d) Time-varying

1.4 (a) Stable
(b) Unstable
(c) Stable
(d) Unstable

1.5 (a) Causal
(b) Causal
(c) Non-causal
(d) Non-causal

1.6 (a) $h[n] = \left(\frac{1}{2}\right)^n u[n]$
(b) $y[n] = \left(\frac{1}{2}\right)^n, \quad 1 \leq n \leq 4$

1.7 (a) $h[n] = \left(\left(\frac{1+\sqrt{2}}{2}\right)^n - \left(\frac{1-\sqrt{2}}{2}\right)^n\right) u[n]$
(b) $y[n] = 2h[n] + h[n-1]$

M. Catak and T. Allahviranloo, *Discrete Random Processes*, Studies in Systems, Decision and Control 644, https://doi.org/10.1007/978-3-032-08630-3

1.8

$$y[n] = \begin{cases} 0 & \text{for } n < -5 \\ -\frac{1}{4}(n+6) & \text{for } -5 \le n < 1 \\ -\frac{3}{2} & \text{for } n \ge 1 \end{cases}$$

1.9 (a) Non-linear
(b) Non-linear
(c) Linear
(d) Linear

1.10 (a) Causal
(b) Non-causal
(c) Causal
(d) Non-causal

1.11 (a) Time-invariant
(b) Time-invariant
(c) Time-varying
(d) Time-invariant

1.12 (a) Unstable
(b) Stable
(c) Stable
(d) Unstable

1.13 (a) Causal
(b) Non-causal
(c) Causal
(d) Causal

1.14 (a) $h[n] = \left(\frac{1}{2}\right)^n u[n]$
(b) $y[n] = \left(\frac{1}{2}\right)^n - \left(\frac{1}{2}\right)^{n-1}, \quad n \ge 1; \quad y[0] = 1$

1.15 (a) Let the roots of the characteristic equation be r_1 and r_2, then

$$h[n] = Ar_1^n + Br_2^n, \quad h[n] = 0 \text{ for } n < 0$$

(b) Use convolution:

$$y[n] = h[n] * u[n]$$

1.16

$$y[n] = \sum_{k=-\infty}^{\infty} (\delta[k] - \delta[k-1]) \cdot 2^{n-k} u[-(n-k)] = 2^n u[-n] - 2^{n-1} u[-n+1]$$

1.17

$$y[n] = \sum_{k=0}^{2} x[k]h[n-k] = \begin{cases} 1, & 0 \le n \le 2 \\ 2, & n = 3, 4 \\ 1, & n = 5 \\ 0, & \text{otherwise} \end{cases}$$

1.18

$$y[n] = \sum_{k=0}^{3} \left(\frac{1}{2}\right)^{n-k} = 15\left(\frac{1}{2}\right)^{n}, \quad n \ge 3$$

1.19 (a) $H(z) = \frac{1+\frac{1}{8}z^{-1}}{1-\frac{1}{4}z^{-1}}$
Inverse Z-transform:

$$h[n] = \left(\frac{1}{4}\right)^{n} u[n] + \frac{1}{8}\left(\frac{1}{4}\right)^{n-1} u[n-1]$$

(b) System is stable since the impulse response is absolutely summable.

1.20 (a) $h[n] = (u[n] - u[n-2]) * (\delta[n] + \delta[n-1]) = \delta[n] + \delta[n-1] + \delta[n-2] + \delta[n-3]$
(b) $y[n] = (\delta[n] + \delta[n-1] + \delta[n-2] + \delta[n-3]) * (\delta[n] - \delta[n-1]) = \delta[n] + \delta[n-1] - \delta[n-2] - \delta[n-3]$

Problems of Chapter 2

2.1

$$h[n] = \left(\frac{1}{2}\right)^{n} u[n] + \left(\frac{1}{4}\right)^{n} u[n]$$

2.2

All poles and zeros must lie inside the unit circle.

2.3

$$x[n] = \left(\frac{1}{6}\left(\frac{1}{2}\right)^{n} + \frac{4}{3}(2)^{n}\right) u[n]$$

2.4

$$x[n] = (-1)^{n+1}\frac{a^{n}}{n}u[n-1]$$

2.5

$$x[n] = \frac{1}{2^n n!} u[n]$$

2.6

$$X(z) = \frac{1 + z^{-1}}{1 - \frac{3}{4} z^{-1}}, \quad \text{ROC: } |z| > \frac{3}{4}$$

2.7

$$X(z) = \frac{z(z - a)}{(z - r_1)(z - r_2)}, \quad \text{ROC: depends on region of convergence}$$

2.8

$$Y(z) = \frac{1}{(1 - az^{-1})^2}, \quad \text{ROC: } |z| > |a|$$

2.9

$$X(z) = \frac{1 - z^{-1}}{1 - \frac{1}{2} z^{-1}}, \quad \text{ROC: } |z| > \frac{1}{2}$$

2.10

$$x[n] = \delta[n] - 2\delta[n - 1] + \delta[n - 2]$$

2.11

$$H(z) = \frac{1 - z^{-1}}{1 + z^{-1}}, \quad \text{ROC: } |z| > 0$$

2.12

$$x[n] = \left(\frac{1}{3}\right)^n u[n] * u[n] = \sum_{k=0}^{n} \left(\frac{1}{3}\right)^k$$

2.13

$$X(z) = \frac{1}{(1 - az^{-1})(1 - bz^{-1})}, \quad \text{ROC: } |z| > \max(|a|, |b|)$$

2.14

$$x[n] = (n + 1)a^n u[n]$$

2.15

$$x[n] = a^n \cos(\omega_0 n) u[n]$$

2.16

$$x[n] = nu[n] \quad \Rightarrow \quad X(z) = \frac{z^{-1}}{(1 - z^{-1})^2}, \quad \text{ROC: } |z| > 1$$

2.17

$$H(z) = \frac{1 - 2z^{-1}}{1 - \frac{1}{2}z^{-1}}, \quad \text{ROC: } |z| > \frac{1}{2}$$

2.18

$$y[n] = \sum_{k=0}^{n} x[k] \quad \text{(accumulator)}$$

2.19

$$H(z) = \frac{z}{z - 1}, \quad \text{ROC: } |z| > 1 \quad \Rightarrow \quad h[n] = u[n]$$

2.20

$$x[n] = \left(\frac{1}{2}\right)^n u[n] + \left(-\frac{1}{3}\right)^n u[n]$$

2.21

$$X(z) = \frac{z}{z - 2}, \quad \text{ROC: } |z| > 2 \quad \Rightarrow \quad x[n] = 2^n u[n]$$

2.22

$$X(z) = \frac{1}{1 - z^{-2}}, \quad \text{ROC: } |z| > 1$$

2.23

$$x[n] = \cos(\omega_0 n)u[n] \quad \Rightarrow \quad X(z) = \frac{1 - \cos(\omega_0)z^{-1}}{1 - 2\cos(\omega_0)z^{-1} + z^{-2}}$$

2.24

$$x[n] = \delta[n] - \delta[n - 2] \quad \Rightarrow \quad X(z) = 1 - z^{-2}$$

2.25

$$H(z) = \frac{1 + z^{-1}}{1 - \frac{1}{2}z^{-1} + \frac{1}{4}z^{-2}}, \quad \text{ROC: } |z| > \frac{1}{2}$$

2.26

$$x[n] = \left(\frac{3}{4}\right)^n u[n] \quad \Rightarrow \quad X(z) = \frac{1}{1 - \frac{3}{4}z^{-1}}, \quad \text{ROC: } |z| > \frac{3}{4}$$

2.27

$$X(z) = \frac{1+z^{-1}}{1-z^{-1}}, \quad \text{ROC: } |z| > 1$$

Problems of Chapter 3

3.1 (a) $X(e^{j\omega}) = \frac{1}{1-\frac{1}{2}e^{-j\omega}}$
(b) $X(e^{j\omega}) = -\frac{1}{3}e^{-j\omega}\frac{1}{1-\frac{1}{3}e^{-j\omega}}$
(c) $X(e^{j\omega}) = 4 - 3e^{-j2\omega}$
(d) $X(e^{j\omega}) = \frac{1-(\frac{1}{2})^9 e^{-j9\omega}}{1-\frac{1}{2}e^{-j\omega}}e^{j2\omega}$

3.2 (a) $X(e^{j\omega}) = 2e^{-j3\omega} + 3e^{-j4\omega}$
(b) $X(e^{j\omega}) = e^{-j\omega} + 4e^{-j2\omega} - e^{-j3\omega}$
(c) $X(e^{j\omega}) = 4e^{-j2.5\times 2\omega}\frac{\sin(3\omega)}{\sin(\omega/2)}$
(d) $X(e^{j\omega}) = \frac{1}{6}\text{rect}(\frac{\omega}{4\pi/3})$ for $|\omega| \le \pi$, periodic with 2π

3.3 (a) $H(e^{j\omega}) = \frac{3+3e^{-j\omega}+3e^{-j2\omega}}{1-3e^{-j\omega}}$
(b) $h[n] = 3(3)^n u[n] + 3(3)^{n-1}u[n-1] + 3(3)^{n-2}u[n-2]$

3.4 (c) $Y(e^{j\omega}) = X\left(e^{j(\omega-2\omega_0)}\right)$

3.5 (a) Continuous and Periodic with a period of 2π

3.6 (c) $X\left(e^{j\omega}\right) = \frac{1}{(1-0.3e^{-j\omega})}$

3.7 (b) $Y\left(e^{j\omega}\right) = X\left(e^{j\omega}\right) - 2X\left(e^{j\omega}\right)e^{-j\omega n_0}$

3.8 (d) $Y\left(e^{j\omega}\right) = 2X\left(e^{j\omega}\right) + j\frac{dX\left(e^{j\omega}\right)}{d\omega}$

3.9 (a) $x[n] = 1$
(b) $x[n] = \delta[n-4]$
(c) $x[n] = \frac{\sin(\omega_c(n+1/2))}{\pi(n+1/2)}$
(d) $x[n] = 2\delta[n] + 4(\delta[n-1] + \delta[n+1]) + 6(\delta[n-2] + \delta[n+2])$

3.10 (a) $X(e^{j\omega}) = \frac{1-a^2}{1-2a\cos(\omega)+a^2}$
(b) $X(e^{j\omega}) = \frac{Ne^{-j\omega(N-1)}}{(1-e^{-j\omega})^2} - \frac{e^{-jN\omega}(1-e^{-jN\omega})}{(1-e^{-j\omega})^2}$
(c) $X(e^{j\omega}) = 1 - e^{-j\omega}$
(d) $X(e^{j\omega}) = \frac{2\pi}{M}\sum_{k=-\infty}^{\infty}\delta(\omega - k\frac{2\pi}{M})$

3.11 (a) $h[n] = (0.5)^n u[n]$
(b) $y[n] = \delta[n] - 0.25\delta[n-1]$

3.12 (a) $Y(e^{j\omega}) = X(e^{-j\omega})$
(b) $Y(e^{j\omega}) = (X(e^{j\omega}))^2$
(c) $Y(e^{j\omega}) = X(e^{j(\omega-\omega_0)})$
(d) $Y(e^{j\omega}) = j\frac{dX(e^{j\omega})}{d\omega}$

3.13 (a) $H(e^{j\omega}) = e^{-j\omega}\frac{\sin(3\omega/2)}{\sin(\omega/2)}$
(b) Magnitude: $|H(e^{j\omega})| = |\frac{\sin(3\omega/2)}{\sin(\omega/2)}|$, Phase: $\angle H(e^{j\omega}) = -\omega + \pi k$ (where k changes when $\sin(3\omega/2)$ or $\sin(\omega/2)$ changes sign)

3.14 $y[n] = \frac{1}{1-0.8e^{-j\omega_0}} e^{j\omega_0 n}u[n] + \frac{0.8}{1-0.8e^{j\omega_0}} \frac{(0.8)^n e^{j\omega_0}}{e^{j\omega_0}-0.8}u[n]$ $y[n] = \frac{1}{1-0.8\cos(\omega_0)+j0.8\sin(\omega_0)}$ $\cos(\omega_0 n)u[n] + ...$

3.15 $x[n] = \frac{1}{2\pi}\int_{-\pi/4}^{\pi/4} e^{-j2\omega}e^{jn\omega}d\omega = \frac{1}{2\pi}\int_{-\pi/4}^{\pi/4} e^{j\omega(n-2)}d\omega = \frac{\sin((n-2)\pi/4)}{\pi(n-2)}$

3.16 $Y(e^{j\omega}) = X(e^{j2\omega})$

3.17 (a) $H(e^{j\omega}) = \frac{1+0.25e^{-j\omega}}{1-0.5e^{-j\omega}}$
(b) $h[n] = (0.5)^n u[n] + 0.25(0.5)^{n-1}u[n-1]$

3.18 $X(e^{j\omega}) = \frac{\sin(2.5\omega)}{\sin(\omega/2)}$, Magnitude: $|X(e^{j\omega})| = |\frac{\sin(2.5\omega)}{\sin(\omega/2)}|$

3.19 $y[n] = 3\delta[n-1] + 2\delta[n-2]$

3.20 Inverse DTFT is $jnx[n]$

Problems of Chapter 4

4.1 The DTFT yields a continuous and periodic frequency spectrum, which cannot be directly stored or processed by finite digital systems. Implementation requires discretization of the frequency axis, leading to the DFT.

4.2 For a finite-length signal $x[n]$ of length N, the DFT $X[k]$ is: Analysis Equation: $X[k] = \sum_{n=0}^{N-1} x[n]W_N^{kn}$, for $k = 0, 1, ..., N-1$ Synthesis Equation: $x[n] = \frac{1}{N}\sum_{k=0}^{N-1} X[k]W_N^{-kn}$, for $n = 0, 1, ..., N-1$

4.3 The twiddle factor $W_N = e^{-j\frac{2\pi}{N}}$. It acts as a complex sinusoidal basis function and is crucial for relating time-domain samples to frequency-domain coefficients in the DFT computation.

4.4 $X[0] = 2 + 1 + 0 - 1 = 2$ $X[1] = 2(1) + 1(-j) + 0(-1) - 1(j) = 2 - 2j$ $X[2] = 2(1) + 1(-1) + 0(1) - 1(-1) = 2$ $X[3] = 2(1) + 1(j) + 0(-1) - 1(-j) = 2 + 2j$ $X[k] = [2, 2-2j, 2, 2+2j]$

4.5 The DFT can be expressed in matrix form as $\mathbf{X} = \mathbf{W_N x}$, where $\mathbf{X}$ is the DFT vector, $\mathbf{x}$ is the input signal vector, and $\mathbf{W_N}$ is the $N \times N$ DFT matrix with elements $[\mathbf{W_N}]_{k,n} = W_N^{kn} = e^{-j\frac{2\pi}{N}kn}$.

4.6 The inverse DFT matrix $\mathbf{W_N^{-1}}$ has elements $[\mathbf{W_N^{-1}}]_{n,k} = \frac{1}{N}W_N^{-kn} = \frac{1}{N}e^{j\frac{2\pi}{N}kn}$. The inverse DFT is computed as $\mathbf{x} = \mathbf{W_N^{-1}X} = \frac{1}{N}\mathbf{W_N}^*\mathbf{X}$, where $*$ denotes the complex conjugate.

4.7 Zero-padding increases the length of the DFT $X[k]$, providing more samples in the frequency domain. These additional samples represent the underlying DTFT evaluated at a finer grid of frequencies. The shape of the DTFT envelope remains the same.

4.8 The DFT can analyze LTI systems by transforming the input $x[n]$ and impulse response $h[n]$ into the frequency domain ($X[k]$ and $H[k]$). Convolution in the time domain corresponds to pointwise multiplication in the frequency domain ($Y[k] = H[k]X[k]$). For linear convolution using DFT, the lengths of $x[n]$ and $h[n]$ must be zero-padded to at least the length of the linear convolution ($N_x + N_h - 1$) to avoid circular convolution effects.

4.9 Zero-padding was necessary to perform linear convolution using the DFT. Without zero-padding, the DFT multiplication would correspond to circular convolution in the time domain, which is generally not the desired output for LTI system analysis. The length of the resulting output $y[n]$ is $N_x + N_h - 1$ in the time domain, and the length of $Y[k]$ in the frequency domain is equal to the zero-padded length used for the DFTs of $x[n]$ and $h[n]$.

4.10 Direct computation of an N-point DFT requires $O(N^2)$ complex multiplications and additions. An N-point FFT algorithm reduces this complexity to $O(N \log_2 N)$ by exploiting symmetries and redundancies in the DFT computation through a divide-and-conquer approach. The FFT is "fast" due to this significantly reduced number of operations, especially for large N.

4.11 The basic FFT algorithm (radix-2) requires the length N of the sequence to be a power of 2 ($N = 2^m$, where m is an integer). If the sequence length is not a power of 2, it is typically zero-padded to the next higher power of 2 before applying the FFT algorithm.

4.12 The FFT algorithm recursively divides the N-point DFT computation into smaller DFTs. It breaks down the DFT of a sequence of length N into two DFTs of length $N/2$ (for even and odd indexed samples). These smaller DFTs are then combined using twiddle factors to produce the N-point DFT. This divide-and-conquer strategy reduces computations by reusing intermediate results.

4.13 In the signal flow graph, the nodes represent the time-domain samples (at the input) or the intermediate and final DFT coefficients. The directed edges represent the flow of signal and are associated with multiplications by twiddle factors or simple additions/subtractions (butterfly operations). The twiddle factors on the edges introduce the necessary phase shifts and scaling for combining the results of smaller DFTs.

4.14 The gain from $x[1]$ to $X[5]$ is determined by tracing all paths from the input node corresponding to $x[1]$ to the output node corresponding to $X[5]$, multiplying the twiddle factors along each path, and summing the results. This complex gain factor is equivalent to the coefficient of $x[1]$ in the direct DFT analysis equation for $X[5]$, which is $W_8^{5\times 1} = e^{-j\frac{2\pi}{8}\times 5} = e^{-j\frac{5\pi}{4}}$.

4.15 Sampling is the process of converting a continuous-time signal $x_a(t)$ into a discrete-time sequence $x[n]$ by taking its values at specific time instants $t = nT_s$, where n is an integer and T_s is the sampling period. The sampling rate or sampling frequency, $f_s = 1/T_s$, is the number of samples taken per unit of time, typically measured in Hertz (Hz) or samples per second.

4.16 The Nyquist-Shannon sampling theorem states that a bandlimited continuous-time signal with a maximum frequency f_{max} can be perfectly reconstructed from its samples if the sampling rate f_s is greater than twice the maximum frequency ($f_s > 2f_{max}$). The Nyquist rate is the minimum sampling rate required for perfect reconstruction, $f_{Nyquist} = 2f_{max}$. Sampling below the Nyquist rate leads to aliasing, making perfect reconstruction impossible.

4.17 Aliasing is the phenomenon where high-frequency components in a continuous-time signal are undersampled (sampled below the Nyquist rate) and appear as lower-frequency components in the discrete-time signal. In the frequency domain, the spectra of the original signal and its shifted replicas (due to sampling) overlap, creating distortion. Aliasing can be avoided by ensuring the sampling rate is at least twice the maximum frequency of the signal or by using an anti-aliasing filter (a low-pass filter) before sampling to limit the signal's bandwidth.

4.18 The 2D DFT of a 2D signal $x[n_1, n_2]$ of size $N_1 \times N_2$ is $X[k_1, k_2]$ of size $N_1 \times N_2$, defined as: Analysis Equation: $X[k_1, k_2] = \sum_{n_1=0}^{N_1-1} \sum_{n_2=0}^{N_2-1} x[n_1, n_2] W_{N_1}^{k_1 n_1} W_{N_2}^{k_2 n_2}$, for $k_1 = 0, ..., N_1 - 1$ and $k_2 = 0, ..., N_2 - 1$ Synthesis Equation: $x[n_1, n_2] = \frac{1}{N_1 N_2} \sum_{k_1=0}^{N_1-1} \sum_{k_2=0}^{N_2-1} X[k_1, k_2] W_{N_1}^{-k_1 n_1} W_{N_2}^{-k_2 n_2}$, for $n_1 = 0, ..., N_1 - 1$ and $n_2 = 0, ..., N_2 - 1$

4.19 The 2D FFT is typically implemented by applying the 1D FFT algorithm separably along each dimension. First, the 1D FFT is computed for each row of the 2D signal. Then, the transpose of the resulting 2D array is taken, and the 1D FFT is computed for each row of this transposed array (which corresponds to the columns of the original data). This approach efficiently computes the 2D DFT with a complexity of $O(N_1 N_2 \log_2(N_1 N_2))$ if N_1 and N_2 are powers of 2.

4.20 *Spectral Analysis:* The FFT is used to decompose a signal into its constituent frequencies, allowing for the identification of dominant frequencies, harmonics, and noise components. This is crucial in audio processing, telecommunications, and vibration analysis.

Convolution and Filtering: The FFT enables efficient implementation of linear convolution by performing multiplication in the frequency domain. This is widely used for filtering signals, such as noise reduction, equalization, and image blurring or sharpening.

Data Compression: In image and audio compression (e.g., JPEG, MP3), the FFT (or related transforms like the DCT) is used to transform the signal into a frequency domain representation where less perceptually significant components can be discarded or quantized with fewer bits, achieving data reduction.

Problems of Chapter 5

5.1 (a) $\mu = 127.5$, $\sigma^2 = \frac{(255-0+1)^2-1}{12} = \frac{256^2-1}{12} = \frac{65535}{12} \approx 5461.25$
(b) $P(X \text{ is divisible by } 5) = \frac{\lfloor 255/5 \rfloor - \lfloor -1/5 \rfloor}{256} = \frac{51-0}{256} = \frac{51}{256} \approx 0.1992$

5.2 (a) $P(Y = 3) = \frac{1}{32} = 0.03125$
(b) $\mu = \frac{m+1}{2}$, $\sigma^2 = \frac{m^2-1}{12}$

5.3 (a) $P(X = 4) = \binom{100}{4}(0.05)^4(0.95)^{96} \approx 0.1781$
(b) $P(X = 4|\text{first 50 unsuccessful}) = P(4 \text{ successes in last } 50) = \binom{50}{4}(0.05)^4 (0.95)^{46} \approx 0.0898$

5.4 (a) $P(X = 8) = \binom{8-1}{3-1}(0.45)^3(1-0.45)^{8-3} = \binom{7}{2}(0.45)^3(0.55)^5 \approx 0.1178$
(b) $\mu = \frac{r}{p} = \frac{3}{0.45} = \frac{20}{3} \approx 6.67$, $\sigma = \sqrt{\frac{r(1-p)}{p^2}} = \sqrt{\frac{3(0.55)}{(0.45)^2}} = \sqrt{\frac{1.65}{0.2025}} \approx 2.8577$

5.5 (a) $\mu = \frac{r}{p} = \frac{5}{0.01} = 500$, $\sigma^2 = \frac{r(1-p)}{p^2} = \frac{5(0.99)}{(0.01)^2} = \frac{4.95}{0.0001} = 49500$
(b) $P(X = 20 \text{ for } r = 2) = \binom{20-1}{2-1}(0.01)^2(0.99)^{19} = 19(0.0001)(0.99)^{19} \approx 0.0016$
(c) $P(X \le 8 \text{ for } r = 5) = \sum_{k=5}^{8} \binom{k-1}{5-1}(0.01)^5(0.99)^{k-5}$. This probability is extremely small, indicating high unreliability for 5 failures within 8 seconds.

5.6 (a) Expected packets in a minute = $3 \times 60 = 180$
(b) $P(X = 120) = \frac{e^{-180}(180)^{120}}{120!} \approx 1.56 \times 10^{-9}$
(c) $P(X \le 240) = \sum_{k=0}^{240} \frac{e^{-180}(180)^k}{k!} \approx 0.9999999995$
(d) An increase in λ increases the expected number of arriving packets and the probability of exceeding the network's capacity, leading to congestion and performance degradation.

5.7 (a) $\lambda_{3 \text{ days}} = 0.6 \times 3 = 1.8$, $P(X = 2) = \frac{e^{-1.8}(1.8)^2}{2!} \approx 0.2678$
(b) $P(X \ge 1) = 1 - P(X = 0) = 1 - \frac{e^{-0.6}(0.6)^0}{0!} = 1 - e^{-0.6} \approx 0.4512$

5.8 (a) $n = 10^6$, $p = 10^{-4}$, $\lambda = np = 100$, $P(X = 100) = \binom{10^6}{100}(10^{-4})^{100} (1-10^{-4})^{10^6-100} \approx 0.0792$
(b) $P(X \le 5) \approx \sum_{k=0}^{5} \frac{e^{-100}(100)^k}{k!} \approx 9.33 \times 10^{-42}$

5.9 (a) Number of possible users = $50 - 10 + 1 = 41$, $P(X = 25) = \frac{1}{41} \approx 0.0244$
(b) $E[X] = \frac{10+50}{2} = 30$

5.10 (a) $r = 5$, $p = 0.2$, $k = 20$, $P(X = 20) = \binom{20-1}{5-1}(0.2)^5(0.8)^{15} \approx 0.01746$
(b) $E[X] = \frac{r}{p} = \frac{5}{0.2} = 25$ hours

5.11 (a) $n = 1000$, $p = 0.01$, $\lambda = np = 10$, $P(X = 15) = \binom{1000}{15}(0.01)^{15}(0.99)^{985} \approx 0.0347$
(b) $P(X \ge 5) = 1 - P(X \le 4) = 1 - \sum_{k=0}^{4} \binom{1000}{k}(0.01)^k(0.99)^{1000-k} \approx 1 - 0.0293 = 0.9707$

5.12 (a) $\lambda_{2 \text{ min}} = 5 \times 2 = 10$, $P(X = 10) = \frac{e^{-10}(10)^{10}}{10!} \approx 0.1251$
(b) $\lambda_{0.5 \text{ min}} = 5 \times 0.5 = 2.5$, $P(X = 0) = \frac{e^{-2.5}(2.5)^0}{0!} = e^{-2.5} \approx 0.0821$

5.13 (a) $n = 100$, $p = 0.02$, $\lambda = np = 2$, $P(X = 3) = \binom{100}{3}(0.02)^3(0.98)^{97} \approx 0.1823$
(b) $P(X > 5) = 1 - P(X \leq 5) = 1 - \sum_{k=0}^{5}\binom{100}{k}(0.02)^k(0.98)^{100-k} \approx 1 - 0.9629 = 0.0371$

5.14 (a) $\lambda_{25\text{ ms}} = 0.8 \times \frac{25}{10} = 2$, $P(X = 0) = \frac{e^{-2}(2)^0}{0!} = e^{-2} \approx 0.1353$
(b) $\lambda_{5\text{ ms}} = 0.8 \times \frac{5}{10} = 0.4$, $P(X \geq 1) = 1 - P(X = 0) = 1 - \frac{e^{-0.4}(0.4)^0}{0!} = 1 - e^{-0.4} \approx 0.3297$

5.15 (a) Number of possible events $= 10 - 0 + 1 = 11$, $P(X = 5) = \frac{1}{11} \approx 0.0909$
(b) Average events per hour $= \frac{0+10}{2} = 5$, Average events in 3 hours $= 5 \times 3 = 15$

5.16 (a) $r = 4$, $p = 0.8$, $k = 6$, $P(X = 6) = \binom{6-1}{4-1}(0.8)^4(0.2)^2 = \binom{5}{3}(0.8)^4(0.2)^2 = 10 \times 0.4096 \times 0.04 = 0.16384$
(b) $P(X \leq 5) = P(X = 4) + P(X = 5) = \binom{4-1}{4-1}(0.8)^4(0.2)^0 + \binom{5-1}{4-1}(0.8)^4 (0.2)^1 = 1 \times 0.4096 \times 1 + 4 \times 0.4096 \times 0.2 = 0.4096 + 0.32768 = 0.73728$

5.17 (a) $n = 1000$, $p = 0.001$, $\lambda = np = 1$, $P(X = 2) = \frac{e^{-1}(1)^2}{2!} = \frac{e^{-1}}{2} \approx 0.1839$
(b) $P(X \geq 1) = 1 - P(X = 0) = 1 - \frac{e^{-1}(1)^0}{0!} = 1 - e^{-1} \approx 0.6321$

5.18 (a) $\lambda = 4$, $P(X = 5) = \frac{e^{-4}(4)^5}{5!} \approx 0.1563$
(b) $\lambda_{0.5\text{ min}} = 4 \times 0.5 = 2$, $P(X < 3) = P(X = 0) + P(X = 1) + P(X = 2) = \frac{e^{-2}(2)^0}{0!} + \frac{e^{-2}(2)^1}{1!} + \frac{e^{-2}(2)^2}{2!} = e^{-2}(1 + 2 + 2) = 5e^{-2} \approx 0.6767$

5.19 (a) Number of possible errors $= 15 - 0 + 1 = 16$, $P(X = 10) = \frac{1}{16} = 0.0625$
(b) $P(5 \leq X \leq 10) = \sum_{k=5}^{10}\frac{1}{16} = \frac{10-5+1}{16} = \frac{6}{16} = \frac{3}{8} = 0.375$

5.20 (a) $r = 3$, $p = 0.7$, $k = 5$, $P(X = 5) = \binom{5-1}{3-1}(0.7)^3(0.3)^2 = \binom{4}{2}(0.7)^3(0.3)^2 = 6 \times 0.343 \times 0.09 = 0.18522$
(b) $E[X] = \frac{r}{p} = \frac{3}{0.7} = \frac{30}{7} \approx 4.2857$

Problems of Chapter 6

6.1 (a) Assuming a typo and $Y[n] = X[n]$: $E[Y[n]] = 3$. If the summation is intended, the expected value is undefined (infinite). (b) Assuming a typo and $Y[n] = X[n]$: $\text{Var}[Y[n]] = 21$. If the summation is intended, the variance will likely diverge.

6.2 (a) Assuming a typo and $Y[n] = 2X[n]$: $E[Y[n]] = 10$. If the summation is intended, the expected value is undefined (infinite). (b) Assuming a typo and $Y[n] = 2X[n]$: $\text{Var}[Y[n]] = 68$. If the summation is intended, the variance will likely diverge.

6.3 The covariance matrix C_X has elements $C_X[i, j] = \text{Cov}(X[i], X[j]) = R_{XX}[i - j] - E[X[i]]E[X[j]]$. Since $E[X[n]] = 0$, $C_X[i, j] = R_{XX}[i - j] = \frac{\sin(\pi(i-j))}{\pi(i-j)}$ with $\frac{\sin(0)}{0} = 1$.

$$C_X = \begin{pmatrix} 1&0&0&0&0&0&0&0&0 \\ 0&1&0&0&0&0&0&0&0 \\ 0&0&1&0&0&0&0&0&0 \\ 0&0&0&1&0&0&0&0&0 \\ 0&0&0&0&1&0&0&0&0 \\ 0&0&0&0&0&1&0&0&0 \\ 0&0&0&0&0&0&1&0&0 \\ 0&0&0&0&0&0&0&1&0 \\ 0&0&0&0&0&0&0&0&1 \end{pmatrix}$$

6.4 (a) $\mu_Y[n] = 0$ (b) $R_{YY}[n_1, n_2] = 2$ (c) Yes, $Y[n]$ is a wide-sense stationary (WSS) process.

6.5 (a) $\mu_Z[n] = \mu_X[n]\mu_Y[n]$ (b) $R_{ZZ}[n_1, n_2] = R_{XX}[n_1, n_2]R_{YY}[n_1, n_2]$

6.6 (a) $\mu_Y[n] = 12$ (b) $R_{XY}[k] = 4(0.5)^{|k|}$ (c) $R_{YY}[k] = 16(0.5)^{|k|}$ (d) Yes, $Y[n]$ is a WSS process.

6.7 (a) $\mu_B[n] = 2$ (b) $R_{AB}[k] = -4e^{-|k|}$ (c) $R_{BB}[k] = 8e^{-|k|}$ (d) Yes, $B[n]$ is a WSS process.

6.8 (a) $\mu_Y[n] = 0$ (b) $R_{XY}[k] = e^{-|k+1|} - e^{-|k|}$ (c) $R_{YY}[k] = 2e^{-|k|} - e^{-|k-1|} - e^{-|k+1|}$ (d) Yes, $Y[n]$ is a WSS process.

6.9 (a) $\mu_Y[n] = 0$ (b) $R_{YY}[k] = \frac{1}{2}(0.5)^{|k|}\cos(\omega k)$ (c) Yes, $Y[n]$ is a WSS process.

6.10 (a) $\mu_Z[n] = 0$ (b) $R_{ZZ}[k] = \frac{3}{2}(0.8)^{|k|}\cos(\omega k)$ (c) Yes, $Z[n]$ is a WSS process.

6.11 (a) $E[Y[n]] = \mu_X + \mu_W$ (b) $R_{YY}[k] = R_{XX}[k] + R_{WW}[k] + \mu_X\mu_W + \mu_W\mu_X = R_{XX}[k] + R_{WW}[k] + 2\mu_X\mu_W$ (c) $R_{XY}[k] = R_{XX}[k] + \mu_X\mu_W$

6.12 (a) $\text{Var}[X[n]] = R_{XX}[0] - (E[X[n]])^2 = A - (E[X[n]])^2$. The mean $E[X[n]]$ is not given, so we leave it in terms of the mean. If the mean is assumed to be 0, then $\text{Var}[X[n]] = A$. (b) $\rho_{XX}[k] = \alpha^{|k|}$

6.13 (a) $E[X[n]] = AE[\cos(\omega_0 n + \Theta)] = A \cdot 0 = 0$ (constant). $R_{XX}[n_1, n_2] = E[A\cos(\omega_0 n_1 + \Theta)A\cos(\omega_0 n_2 + \Theta)] = A^2 E[\cos(\omega_0 n_1 + \Theta)\cos(\omega_0 n_2 + \Theta)] = \frac{A^2}{2}\cos(\omega_0(n_1 - n_2))$. This depends only on the time difference $n_1 - n_2$. Thus, $X[n]$ is WSS. (b) $R_{XX}[k] = \frac{A^2}{2}\cos(\omega_0 k)$

6.14 (a) $R_{YY}[k] = R_{XX}[k] * h[k] * h[-k] = R_{XX}[k] * (a^k u[k]) * (a^{-k}u[-k])$. Since $X[n]$ is zero-mean, $R_{XX}[k] = E[X[n]X[n+k]]$. $R_{YY}[k] = \sum_{m=-\infty}^{\infty}\sum_{n=-\infty}^{\infty} R_{XX}[k - m + n]h[m]h[n]$ $R_{YY}[k] = \sum_{m=0}^{\infty}\sum_{n=0}^{\infty} R_{XX}[k - m + n]a^m a^n$ (b) Yes, if $X[n]$ is WSS and the LTI system is time-invariant, then the output $Y[n]$ is also WSS.

6.15 (a) $R_{ZZ}[k] = R_{XX}[k] + R_{YY}[k] - R_{XY}[k] - R_{YX}[k] = R_{XX}[k] + R_{YY}[k] - 2R_{XY}[k]$ (since $X[n]$ and $Y[n]$ are jointly WSS, $R_{YX}[k] = R_{XY}[-k]$, but if we assume R_{XY} is even, then $R_{YX}[k] = R_{XY}[k]$). (b) $R_{XZ}[k] = E[X[n](X[n+k] - Y[n+k])] = E[X[n]X[n+k]] - E[X[n]Y[n+k]] = R_{XX}[k] - R_{XY}[k]$

6.16 (a) $E[Y[n]] = E[X[n] + 2X[n-1] + X[n-2]] = E[X[n]] + 2E[X[n-1]] + E[X[n-2]] = \mu_X + 2\mu_X + \mu_X = 4\mu_X$ (b) $R_{YY}[k] = E[Y[n]Y[n+k]] = E[(X[n] + 2X[n-1] + X[n-2])(X[n+k] + 2X[n+k-1] + X[n+k-2])]$ $R_{YY}[k] = R_{XX}[k] + 2R_{XX}[k-1] + R_{XX}[k-2] + 2R_{XX}[k+1] + 4R_{XX}[k] + 2R_{XX}[k-1] + R_{XX}[k+2] + 2R_{XX}[k+1] + R_{XX}[k]$ $R_{YY}[k] = 6R_{XX}[k] + 4R_{XX}[k-1] + 4R_{XX}[k+1] + R_{XX}[k-2] + R_{XX}[k+2]$

6.17 (a) $R_{YY}[k] = E[(X[n] - X[n-1])(X[n+k] - X[n+k-1])] = E[X[n]X[n+k]] - E[X[n]X[n+k-1]] - E[X[n-1]X[n+k]] + E[X[n-1]X[n+k-1]]$ $R_{YY}[k] = R_{XX}[k] - R_{XX}[k-1] - R_{XX}[k+1] + R_{XX}[k] = \sigma^2\delta[k] - \sigma^2\delta[k-1] - \sigma^2\delta[k+1] + \sigma^2\delta[k] = 2\sigma^2\delta[k] - \sigma^2\delta[k-1] - \sigma^2\delta[k+1]$
(b) $R_{XY}[k] = E[X[n](X[n+k] - X[n+k-1])] = E[X[n]X[n+k]] - E[X[n]X[n+k-1]] = R_{XX}[k] - R_{XX}[k-1] = \sigma^2\delta[k] - \sigma^2\delta[k-1]$

6.18 (a) $R_{YY}[k] = E[(aX[n] + b)(aX[n+k] + b)] = a^2E[X[n]X[n+k]] + abE[X[n]] + abE[X[n+k]] + b^2 = a^2R_{XX}[k] + ab\mu_X + ab\mu_X + b^2 = a^2(0.9)^{|k|} + 2ab\mu_X + b^2$ (b) For $Y[n]$ to be zero-mean, $E[Y[n]] = E[aX[n] + b] = a\mu_X + b = 0$, so $b = -a\mu_X$. The value of a can be any constant.

6.19 (a) $E[X[n]] = E[A + Bn] = E[A] + nE[B] = \mu_A + n\mu_B$ (b) $R_{XX}[n_1, n_2] = E[(A + Bn_1)(A + Bn_2)] = E[A^2 + ABn_2 + ABn_1 + B^2n_1n_2] = E[A^2] + n_2E[AB] + n_1E[AB] + n_1n_2E[B^2]$ Since A and B are uncorrelated, $E[AB] = E[A]E[B] = \mu_A\mu_B$. Also, $E[A^2] = \mathrm{Var}(A) + (E[A])^2 = \sigma_A^2 + \mu_A^2$, and $E[B^2] = \mathrm{Var}(B) + (E[B])^2 = \sigma_B^2 + \mu_B^2$. $R_{XX}[n_1, n_2] = \sigma_A^2 + \mu_A^2 + n_2\mu_A\mu_B + n_1\mu_A\mu_B + n_1n_2(\sigma_B^2 + \mu_B^2)$ (c) No, $X[n]$ is not WSS because the mean function $E[X[n]] = \mu_A + n\mu_B$ depends on time n, and the autocorrelation function $R_{XX}[n_1, n_2]$ depends on the specific times n_1 and n_2, not just their difference.

6.20 (a) $R_{YY}[k] = E[Y[n]Y[n+k]] = E[X[-n]X[-(n+k)]] = E[X[-n]X[-n-k]]$ Let $m = -n$, then $R_{YY}[k] = E[X[m]X[m-k]] = R_{XX}[-k]$. Since $R_{XX}[k]$ is an even function for a WSS process, $R_{YY}[k] = R_{XX}[k]$. (b) $R_{XY}[n_1, n_2] = E[X[n_1]Y[n_2]] = E[X[n_1]X[-n_2]]$ (c) Yes, $Y[n]$ is WSS because $E[Y[n]] = E[X[-n]] = \mu_X$ (constant) and $R_{YY}[k] = R_{XX}[k]$ (depends only on the time difference).

Problems of Chapter 7

7.1

(a) The variance of the process X[n] is given by the autocovariance at lag 0: $\mathrm{Var}(X[n]) = R_X[0] = 5 \cdot (0.7)^{|0|} = 5$.
(b) The autocorrelation matrix R_X for $k = -2, -1, 0, 1, 2$ is:

$$R_X = \begin{bmatrix} R_X[0] & R_X[1] & R_X[2] \\ R_X[-1] & R_X[0] & R_X[1] \\ R_X[-2] & R_X[-1] & R_X[0] \end{bmatrix} = \begin{bmatrix} 5 & 3.5 & 2.45 \\ 3.5 & 5 & 3.5 \\ 2.45 & 3.5 & 5 \end{bmatrix}$$

7.2

(a) The cross-correlation function $R_{YX}[k]$ is given by $R_{YX}[k] = R_{XY}[-k] = 3 \cdot (0.5)^{-k}u[-k]$.
(b) Since the processes have zero means, the cross-covariance function $C_{XY}[k]$ is equal to the cross-correlation function: $C_{XY}[k] = R_{XY}[k] = 3 \cdot (0.5)^k u[k]$.

7.3

(a) The power spectral density (PSD) $S_X(e^{j\omega})$ is the DTFT of $R_X[k]$:

$$S_X(e^{j\omega}) = \sum_{k=-\infty}^{\infty} R_X[k]e^{-j\omega k} = 8 \sum_{k=-\infty}^{\infty} \cos(0.2\pi k)e^{-j\omega k} + 3$$

Using the property that $\cos(x) = \frac{e^{jx}+e^{-jx}}{2}$, we get:

$$S_X(e^{j\omega}) = 4\pi [\delta(\omega - 0.2\pi) + \delta(\omega + 0.2\pi)] + 3, \quad -\pi \le \omega \le \pi$$

(b) The total average power of the process is $R_X[0] = 8\cos(0) + 3 = 11$.

7.4 The autocovariance function is $C_X[k] = R_X[k] - \mu_X^2 = 2 + (0.8)^{|k|} - 1 = 1 + (0.8)^{|k|}$.

$$C_X = \begin{bmatrix} C_X[0] & C_X[1] & C_X[2] \\ C_X[-1] & C_X[0] & C_X[1] \\ C_X[-2] & C_X[-1] & C_X[0] \end{bmatrix} = \begin{bmatrix} 2 & 1.8 & 1.64 \\ 1.8 & 2 & 1.8 \\ 1.64 & 1.8 & 2 \end{bmatrix}$$

7.5

$$SNR_{dB} = 10\log_{10}\left(\frac{P_X}{\sigma^2}\right) = 10\log_{10}\left(\frac{10}{0.5}\right) = 10\log_{10}(20) \approx 13.01 \text{ dB}$$

7.6

$$K_k = P_{k|k-1}H^T(HP_{k|k-1}H^T + R)^{-1} = \begin{bmatrix} 3 & 1 \\ 1 & 2 \end{bmatrix}\begin{bmatrix} 1 \\ 1 \end{bmatrix}\left(\begin{bmatrix} 1 & 1 \end{bmatrix}\begin{bmatrix} 3 & 1 \\ 1 & 2 \end{bmatrix}\begin{bmatrix} 1 \\ 1 \end{bmatrix} + 1\right)^{-1}$$

$$= \begin{bmatrix} 4 \\ 3 \end{bmatrix}(3 + 1 + 1 + 2 + 1)^{-1} = \begin{bmatrix} 4 \\ 3 \end{bmatrix}(8)^{-1} = \begin{bmatrix} 0.5 \\ 0.375 \end{bmatrix}$$

7.7

$$R_Y[k] = E[Y[n]Y[n+k]] = E[(2X[n] + X[n-1])(2X[n+k] + X[n+k-1])]$$

$$= 4R_X[k] + 2R_X[k+1] + 2R_X[k-1] + R_X[k] = 5R_X[k] + 2R_X[k+1] + 2R_X[k-1]$$

7.8

(a) The variance of the process is $\text{Var}(X[n]) = R_X[0] = A\alpha^0 = A$.
(b) The autocorrelation coefficient is $\rho_X[k] = \frac{R_X[k]}{R_X[0]} = \frac{A\alpha^{|k|}}{A} = \alpha^{|k|}$.

7.9 The covariance kernel is $k[n, n'] = \sigma_f^2 \exp\left(-\frac{(n-n')^2}{2\ell^2}\right) = 9\exp\left(-\frac{(n-n')^2}{8}\right)$. The covariance matrix K is:

$$K = \begin{bmatrix} k[1,1] & k[1,2] & k[1,3] & k[1,4] & k[1,5] \\ k[2,1] & k[2,2] & k[2,3] & k[2,4] & k[2,5] \\ k[3,1] & k[3,2] & k[3,3] & k[3,4] & k[3,5] \\ k[4,1] & k[4,2] & k[4,3] & k[4,4] & k[4,5] \\ k[5,1] & k[5,2] & k[5,3] & k[5,4] & k[5,5] \end{bmatrix} = 9\begin{bmatrix} 1 & e^{-1/8} & e^{-1} & e^{-9/8} & e^{-2} \\ e^{-1/8} & 1 & e^{-1/8} & e^{-1} & e^{-9/8} \\ e^{-1} & e^{-1/8} & 1 & e^{-1/8} & e^{-1} \\ e^{-9/8} & e^{-1} & e^{-1/8} & 1 & e^{-1/8} \\ e^{-2} & e^{-9/8} & e^{-1} & e^{-1/8} & 1 \end{bmatrix}$$

7.10 $S_X(e^{j\omega}) = 2 + 4\cos(\omega) + \cos(2\omega)$. $S_X(e^{j\omega}) = 2 + 2(e^{j\omega} + e^{-j\omega}) + \frac{1}{2}(e^{j2\omega} + e^{-j2\omega})$. Therefore, $R_X[0] = 2$, $R_X[1] = R_X[-1] = 2$, $R_X[2] = R_X[-2] = 0.5$, and $R_X[k] = 0$ for $|k| > 2$.

7.11 $C_Y[k] = E[(Y[n] - E[Y[n]])(Y[n+k] - E[Y[n+k]])] = E[(X[n] + X[n-1])(X[n+k] + X[n+k-1])] = E[X[n]X[n+k] + X[n]X[n+k-1] + X[n-1]X[n+k] + X[n-1]X[n+k-1]] = C_X[k] + C_X[k+1] + C_X[k-1] + C_X[k] = 2C_X[k] + C_X[k+1] + C_X[k-1] = 2\sigma^2\delta[k] + \sigma^2\delta[k+1] + \sigma^2 \delta[k-1]$

7.12 Since X[n] and Y[n] are independent, $R_{XY}[k] = R_{YX}[k] = 0$. $R_Z[k] = E[Z[n] Z[n+k]] = E[(X[n] + Y[n])(X[n+k] + Y[n+k])] = E[X[n]X[n+k]] + E[X[n]Y[n+k]] + E[Y[n]X[n+k]] + E[Y[n]Y[n+k]] = R_X[k] + R_Y[k] = (0.5)^{|k|} + 2\delta[k]$

7.13 $h[-k] = \{0.25, 0.5, 1\}$ $R_Y[k] = R_X[k] * h[k] * h[-k] = \delta[k] * \{1, 0.5, 0.25\} * \{0.25, 0.5, 1\} = \{1, 0.5, 0.25\} * \{0.25, 0.5, 1\} = \{0.25, 0.5 + 0.125, 1 + 0.25 + 0.25, 0.5 + 0.125, 0.25\} = \{0.25, 0.625, 1.5, 0.625, 0.25\}$

7.14 The power in the process is given by $R_X[0] = 2(0.8)^0 = 2$.

7.15 $P_e = Q\left(\sqrt{\frac{2\times 8}{0.4}}\right) = Q(\sqrt{40}) = Q(6.32) \approx 10^{-9}$

7.16 The correlation coefficient ρ_{X_1,X_2} is given by $\frac{Cov(X_1,X_2)}{\sigma_{X_1}\sigma_{X_2}}$. From the covariance matrix, $Cov(X_1, X_2) = 1, \sigma_{X_1} = \sqrt{2}$, and $\sigma_{X_2} = \sqrt{3}$. $\rho_{X_1,X_2} = \frac{1}{\sqrt{2}\sqrt{3}} = \frac{1}{\sqrt{6}} \approx 0.408$

7.17 $S_X(e^{j\omega}) = \sum_{k=-\infty}^{\infty} R_X[k]e^{-j\omega k} = \sum_{k=-\infty}^{\infty}(4\delta[k] + (0.6)^{|k|})e^{-j\omega k}$
$= 4 + \sum_{k=-\infty}^{\infty}(0.6)^{|k|}e^{-j\omega k} = 4 + \frac{1-0.6^2}{1-2(0.6)\cos(\omega)+0.6^2} = 4 + \frac{0.64}{1.36-1.2\cos(\omega)}$

7.18 Since the process is Gaussian and has a mean, and if we assume it is also ergodic in the mean, the time average of a long realization will converge to the ensemble mean. Therefore, you would take a long sequence of wave height measurements and calculate the sample mean.

7.19 If R is underestimated, the filter trusts the measurements too much, leading to noisy estimates and potentially instability. If R is overestimated, the filter trusts the model too much, resulting in slow response to changes and potentially biased estimates.

7.20 The squared exponential kernel captures smooth trends, while a periodic kernel can capture seasonality. The choice depends on the expected patterns in the data. An inappropriate kernel can lead to poor predictions.

Problems of Chapter 8

8.1 (b) (i) Irreducible (ii) Aperiodic

(c) Yes, this Markov chain is ergodic because it is irreducible and aperiodic.

(d) $\mathbf{P}^2 = \begin{bmatrix} 0.48\ 0.32\ 0.20 \\ 0.33\ 0.41\ 0.26 \\ 0.21\ 0.40\ 0.39 \end{bmatrix}$ $\mathbf{P}^3 = \begin{bmatrix} 0.414\ 0.356\ 0.230 \\ 0.345\ 0.391\ 0.264 \\ 0.271\ 0.386\ 0.343 \end{bmatrix}$ $\mathbf{P}^n_{ij}$ represents the probability of being in state j after n steps, starting from state i.

(e) To find the stationary distribution $\boldsymbol{\pi} = [\pi_1, \pi_2, \pi_3]$, we solve $\boldsymbol{\pi}\mathbf{P} = \boldsymbol{\pi}$ and $\sum_{i=1}^{3} \pi_i = 1$. Solving this system of equations yields $\boldsymbol{\pi} = [0.3056, 0.3889, 0.3056]$.

(f) $\mathbf{a}[2] = \mathbf{a}[0]\mathbf{P}^2 = [0.5, 0.3, 0.2] \begin{bmatrix} 0.48\ 0.32\ 0.20 \\ 0.33\ 0.41\ 0.26 \\ 0.21\ 0.40\ 0.39 \end{bmatrix} = [0.381, 0.363, 0.256]$.

(g) Yes, the limiting distribution exists and is equal to the stationary distribution because the chain is ergodic. The limiting distribution is $\boldsymbol{\pi} = [0.3056, 0.3889, 0.3056]$.

8.2 (b) The chain moves in a cycle $A \to B \to C \to D \to A$, so the period of each state is 4. The chain is periodic with period 4.

(c) Yes, the chain is irreducible because it is possible to reach any state from any other state (following the cycle).

(d) All states are recurrent because once the chain enters a state, it will eventually return to that state (due to the cyclical nature). There are no transient states.

(e) A limiting distribution does not exist for this chain because it is periodic. The state probabilities will cycle through values and not converge to a single distribution as $n \to \infty$. However, a stationary distribution $\boldsymbol{\pi}$ does exist and is $\boldsymbol{\pi} = [1/4, 1/4, 1/4, 1/4]$.

8.3 (a) States $\{1, 2\}$ form a recurrent class, and states $\{3, 4\}$ form another recurrent class. There are no transient states.
(b) No absorbing states exist.
(c) Yes, a stationary distribution exists (in fact, a family of them). Examples include $[1/3, 2/3, 0, 0]$ and $[0, 0, 3/5, 2/5]$.

8.4 (a) $P(\text{at position 2 after 3 steps}) = 0.125$. (b) $E[\text{position after } n \text{ steps}] = 0$. (c) The symmetric random walk is recurrent in 1D, so it will return to the origin with probability 1.

8.5 (a) $P(\text{at position 2 after 3 steps}) = 0$. (b) $E[\text{position after } n \text{ steps}] = 0.2n$. (c) The nonsymmetric random walk is transient in 1D, so the probability of returning to the origin as $n \to \infty$ is less than 1. The expected position drifts away from the origin.

8.6 (b) $P_{i,i+1} = p(1-q)$, $P_{i,i} = pq + (1-p)(1-q)$ for $i > 0$, $P_{0,0} = 1 - p$, $P_{i,i-1} = (1-p)q$ for $i > 0$. (c) $P(X_3 = 2|X_0 = 0) = 0.03696$. (d) $\pi_i = (1-\rho)\rho^i$, where $\rho = p/q = 2/3$. $\pi_i = (1/3)(2/3)^i$. (e) Average number of packets = 2. (f) Average time in buffer = 10 slots.

8.7 (b) $P(\text{Busy after 2 transitions} \mid \text{starts Idle}) = 0.46$. (c) $\boldsymbol{\pi} = [0.3448, 0.4138, 0.2414]$. (d) Yes, the chain is ergodic (irreducible and aperiodic).

8.8 (b) $\mathbf{P}^2 = \begin{bmatrix} 0.35 & 0.55 & 0.10 \\ 0.22 & 0.46 & 0.32 \\ 0.08 & 0.48 & 0.44 \end{bmatrix}$. (c) $\boldsymbol{\pi} = [4/15, 6/15, 5/15] \approx [0.2667, 0.4000, 0.3333]$. (d) Yes, the chain is irreducible and aperiodic.

8.9 (a) $\mathbf{P} = \begin{bmatrix} 1-p & p \\ q & 1-q \end{bmatrix}$. (c) $\boldsymbol{\pi} = \left[\frac{q}{p+q}, \frac{p}{p+q}\right]$.

8.10 (b) $P(\text{Decrease after 2 periods} \mid \text{starts Increasing}) = 0.24$. (c) $\boldsymbol{\pi} = [11/30, 13/30, 6/30] \approx [0.3667, 0.4333, 0.2000]$.

8.11 (b) All states have a period of 2. (c) Irreducible: Yes. Aperiodic: No.

8.12 (b) State K (Checkout Page) is an absorbing state. (c) The stationary distribution is $\boldsymbol{\pi} = [0, 0, 0, 1]$, indicating that eventually all user sessions end at the Checkout Page.

8.13 (b) Probability of eventual failure starting from Active = 1. (c) No, the chain is not ergodic (not irreducible due to the absorbing state).

8.14 (b) $\mathbf{P}^2 = \begin{bmatrix} 0.52 & 0.16 & 0.32 \\ 0.16 & 0.52 & 0.32 \\ 0.40 & 0.32 & 0.36 \end{bmatrix}$, $\mathbf{P}^3 = \begin{bmatrix} 0.224 & 0.440 & 0.336 \\ 0.440 & 0.224 & 0.336 \\ 0.336 & 0.384 & 0.280 \end{bmatrix}$. (c) $\boldsymbol{\pi} = [0.3, 0.3, 0.4]$.

8.15 (b) State D (Delivered) is an absorbing state. (c) The stationary distribution is $\boldsymbol{\pi} = [0, 0, 0, 1]$, signifying that all packets eventually reach the Delivered state.

8.16 (b) $P(\text{Rainy in 2 days} \mid \text{starts Sunny}) = 0.13$. (c) $\boldsymbol{\pi} = [0.4, 0.4, 0.2]$.

8.17 (b) All states have a period of 2. (c) Irreducible: Yes. Aperiodic: No.

8.18 (b) $\boldsymbol{\pi} = [31/87, 34/87, 22/87] \approx [0.3563, 0.3908, 0.2529]$. (c) The stationary distribution represents the long-term proportion of time the system spends in each state.

8.19 (b) State U (Unavailable) is an absorbing state. (c) Probability of eventually becoming Unavailable starting from Local = 1.

8.20 (b) Irreducible: Yes. Aperiodic: Yes. (c) $\boldsymbol{\pi} = [1/4, 1/4, 1/4, 1/4]$.

Problems of Chapter 9

9.1 (a) $P(X = 5) = \frac{4^5 e^{-4}}{5!} \approx 0.1563$
(b) $P(X \geq 2) = 1 - P(0) - P(1) = 1 - e^{-4} - 4e^{-4} \approx 1 - 0.0183 - 0.0733 = 0.9084$
(c) In 30 minutes, $\lambda = 2$, so $P(X = 0) = e^{-2} \approx 0.1353$

9.2 (a) $\mathbb{E}[X] = 20 \times 5 = 100$
(b) $\text{Var}(X) = 20 \times 10 = 200$
(c) For 12 hours, $\lambda = 10$, so $\sigma = \sqrt{10} \approx 3.1623$

9.3 (a) $\lambda = \lambda_1 + \lambda_2 = 10 + 15 = 25$
(b) $P(X = 20) = \frac{25^{20} e^{-25}}{20!} \approx 0.0512$
(c) $\mathbb{E}[X] = 25 \times 3 = 75$

9.4 (a) $\lambda_A = 0.5 \times 0.4 = 0.2$
(b) $\lambda_B = 0.5 \times 0.6 = 0.3$, $P(X = 1) = 0.3e^{-0.3} \approx 0.2222$
(c) $\mathbb{E}[X] = 10 \times 0.2 = 2$

9.5 (a) $P(T = 1) = 1 - e^{-2} \approx 0.8647$
(b) $P(T > 2) = e^{-2\times 2} = e^{-4} \approx 0.0183$
(c) $P(T = 3) = e^{-2(2)}(1 - e^{-2}) \approx 0.0183 \times 0.8647 \approx 0.0158$

9.6 (a) $P(X = 3|X \geq 1) = \frac{P(3)}{1-P(0)} = \frac{e^{-5}\frac{5^3}{3!}}{1-e^{-5}} \approx \frac{0.1404}{0.9933} \approx 0.1414$
(b) Binomial with parameters $n = 4$, $p = 0.5$: $P(X = 2) = \binom{4}{2}(0.5)^2(0.5)^2 = 0.375$

9.7 (a) $\lambda = 1$, $P(2) = \frac{1^2 e^{-1}}{2!} = \frac{e^{-1}}{2} \approx 0.1839$
(b) Binomial with $n = 10$, $p = 0.1$, $P(2) = \binom{10}{2}(0.1)^2(0.9)^8 \approx 0.1937$
(c) The Poisson is suitable for large n, small p, and moderate np.

9.8 (a) $\mathbb{E}[S_2] = \lambda t \mu = 4 \times 2 \times 20 = 160$
(b) $\text{Var}(S_2) = \lambda t(\sigma^2 + \mu^2) = 8(100 + 400) = 4000$

9.9 To extend to a non-homogeneous Poisson process, let $\lambda(t)$ vary with time. In each time unit, model arrivals as Poisson($\lambda(t)$). Expected values become integrals over time: $\mathbb{E}[X(t)] = \int_0^t \lambda(s)ds$.

9.10 Generate Poisson counts with probability $P(k) = \frac{\lambda^k e^{-\lambda}}{k!}$ or simulate inter-arrival times from a geometric distribution.
Pseudocode:

```
for t in range(0, T):
    u = random()
    if u < Poisson probability:
        event occurs
```

9.11 (a) $\lambda = 14$, $P(10) = \frac{14^{10}e^{-14}}{10!} \approx 0.1049$
(b) $\lambda = \frac{7}{6}$, $P(\geq 1) = 1 - e^{-\frac{7}{6}} \approx 0.8122$
(c) $P(3|X \geq 1) = \frac{P(3)}{1-P(0)} \approx \frac{0.0596}{0.9991} \approx 0.0597$

9.12 (a) $\lambda = 0.2$, $P(0) = e^{-0.2} \approx 0.8187$
(b) $\lambda = 4$, $P(X > 5) = 1 - \sum_{k=0}^{5} \frac{4^k e^{-4}}{k!} \approx 1 - 0.7851 = 0.2149$
(c) $P(0) = e^{-0.002N} \geq 0.99 \Rightarrow N \leq \frac{-\ln(0.99)}{0.002} \approx 5.01$

9.13 (a) $P(X_t = k) = \frac{(\lambda t)^k e^{-\lambda t}}{k!}$
(b) Mean = λt, Variance = λt
(c) Independent increments: $X_{t+s} - X_s$ is independent of X_u for $u \leq s$

9.14 (a) $X_t + Y_t \sim \text{Poisson}(\lambda_1 + \lambda_2)$
(b) $P(X_t = k|X_t + Y_t = n) = \binom{n}{k}\left(\frac{\lambda_1}{\lambda_1+\lambda_2}\right)^k\left(\frac{\lambda_2}{\lambda_1+\lambda_2}\right)^{n-k}$
(c) This is a binomial distribution: conditional distribution given sum is known.

9.15 (a) $P(X > 15) = 1 - \sum_{k=0}^{15} \frac{10^k e^{-10}}{k!} \approx 0.0486$
(b) Let $p = 0.0486$, $P(Y \geq 2)$ where $Y \sim \text{Bin}(5, 0.0486)$, $P(Y \geq 2) = 1 - P(0) - P(1) \approx 1 - 0.784 - 0.199 = 0.017$
(c) $P(X > 20) \approx 1 - \sum_{k=0}^{20} \frac{10^k e^{-10}}{k!} \approx 0.0016$

9.16 (a) $\lambda = 4$, $P(4) = \frac{4^4 e^{-4}}{4!} \approx 0.1954$
(b) $\lambda = 13$, $P(0) = e^{-13} \approx 2.2 \times 10^{-6}$
(c) $T_5 \sim \text{Negative Binomial}(5, e^{-\lambda})$, or more precisely, $P(T_5 = 10) = \binom{9}{4}(1 - e^{-\lambda})^5 e^{-5\lambda}$

9.17 (a) $P(D > S) = 1 - \sum_{k=0}^{S} \frac{\lambda^k e^{-\lambda}}{k!}$
(b) For $\lambda = 5$, $S = 7$: $P(D > 7) = 1 - \sum_{k=0}^{7} \frac{5^k e^{-5}}{k!} \approx 0.1334$
(c) Find smallest S such that $P(D > S) < 0.01$: $S = 11$

9.18 (a) $T_1 \sim \text{Geometric}(p = 1 - e^{-\lambda})$
(b) $\mathbb{E}[T_k] = \frac{k}{1-e^{-\lambda}}$
(c) $\text{Var}(T_k) = \frac{ke^{-\lambda}}{(1-e^{-\lambda})^2}$

9.19 (a) $\lambda = 1.4$, $P(3) = \frac{1.4^3 e^{-1.4}}{3!} \approx 0.1911$
(b) $\lambda = 0.2$, $P(\geq 1) = 1 - e^{-0.2} \approx 0.1813$
(c) Expected = $0.2 \times 30 = 6$

9.20 To test if data is from a discrete-time Poisson process: - Check if counts per interval are Poisson distributed: use chi-squared goodness-of-fit test. - Check for independent increments: autocorrelation tests or runs tests. - Use Q-Q plot vs theoretical Poisson.

Assumptions: – Events occur independently.
– Rate is constant.
– No simultaneous events in the same time unit.

Problems of Chapter 10

10.1

(a) Given $X_0 = 0,\ X_1 = \alpha X_0 + \beta + \sigma_d \epsilon_1 = 0 + 1 + 0.3\epsilon_1$. Taking expectation: $\mathbb{E}[X_1] = 1$.
(b) Since $\epsilon_1 \sim \mathcal{N}(0, 1)$ and independent: $\mathrm{Var}(X_1) = \sigma_d^2 = 0.09$.
(c) $X_2 = \alpha X_1 + \beta + \sigma_d \epsilon_2 = 0.8 \cdot 2 + 1 + 0.3\epsilon_2$. So, $\mathbb{E}[X_2|X_1 = 2] = 0.8 \cdot 2 + 1 = 2.6$.

10.2

(a) Long-term mean: $\mu = \frac{\beta}{1-\alpha} = \frac{0.5}{1-0.9} = 5$.
(b) Stationary variance: $\sigma_X^2 = \frac{\sigma_d^2}{1-\alpha^2} = \frac{0.04}{1-0.81} = \frac{0.04}{0.19} \approx 0.2105$.
(c) $X_{n+1} \sim \mathcal{N}(0.9 \cdot 3 + 0.5, 0.2^2) = \mathcal{N}(3.2, 0.04)$.

10.3

(a) $\alpha = e^{-\theta \Delta t} = e^{-0.5 \cdot 0.2} = e^{-0.1} \approx 0.9048$.
$\beta = \mu(1 - \alpha) = 10(1 - 0.9048) = 10(0.0952) = 0.952$.
(b) $\sigma_d^2 = \sigma^2 \frac{1-\alpha^2}{2\theta} = 4 \cdot \frac{1-(0.9048)^2}{1} = 4 \cdot (1 - 0.8187) = 4 \cdot 0.1813 = 0.7252$.
Thus, $\sigma_d \approx \sqrt{0.7252} \approx 0.8516$.

10.4

(a) Process-2 with $\alpha = 0.95$ exhibits stronger persistence since it retains more of its previous state.
(b) Process-1 with $\alpha = 0.7$ will revert to the mean faster due to lower persistence.

10.5

(a) Long-term mean: $\mu = \frac{\beta}{1-\alpha} = \frac{0.005}{0.02} = 0.25$.
(b) $\mathbb{E}[r_{n+1}|r_n = 0.03] = 0.98 \cdot 0.03 + 0.005 = 0.0294 + 0.005 = 0.0344$.

10.6 Using $\alpha = e^{-\theta \Delta t}$ and $\sigma_d^2 = \sigma^2 \frac{1-\alpha^2}{2\theta}$, we substitute into the stationary variance:

$$\sigma_X^2 = \frac{\sigma_d^2}{1 - \alpha^2} = \frac{\sigma^2}{2\theta}$$

This confirms the equivalence of the continuous and discrete-time forms under the transformation.

10.7

$$X_1 = \alpha X_0 + \beta + \sigma_d \epsilon_1 = 0.6 \cdot 2 - 1 + 0.4 \cdot 1.5 = 1.2 - 1 + 0.6 = 0.8$$

10.8 An OU process is appropriate for commodities due to mean reversion driven by supply/demand forces. Parameters relate as: - μ: long-run average price (equilibrium), - θ: speed of reversion (reflects regulatory/stabilization), - σ: market volatility (economic shocks).

10.9 As Δt increases: - $\alpha = e^{-\theta \Delta t}$ decreases, - $\beta = \mu(1-\alpha)$ increases, - $\sigma_d^2 = \sigma^2 \frac{1-\alpha^2}{2\theta}$ increases (more diffusion per step).

10.10 With strong mean reversion (α near 0), the process quickly returns to the mean and fluctuates tightly. With weak mean reversion ($\alpha \approx 1$), it drifts slowly and retains memory of past values longer.

10.11

(a) $\mathbb{E}[X_{n+1}|X_n = x] = 0.7x - 0.2$.
(b) $\epsilon_n \sim \mathcal{N}(0, 1)$, so noise is $\mathcal{N}(0, 0.25)$.
(c) $X_1 = 0.7(-1) - 0.2 + 0.5\epsilon_1 \Rightarrow \mathbb{E}[X_1] = -0.7 - 0.2 = -0.9$
$X_2 = 0.7X_1 - 0.2 + 0.5\epsilon_2 \Rightarrow \mathbb{E}[X_2] = 0.7(-0.9) - 0.2 = -0.83$.

10.12 The long-term mean $\mu = \frac{\beta}{1-\alpha} \Rightarrow \beta = \mu(1-\alpha)$.

10.13

$$\alpha = e^{-\gamma \Delta t}, \quad \sigma_d^2 = \eta^2 \cdot \frac{1-\alpha^2}{2\gamma}$$

Thus, $\sigma_d = \eta\sqrt{\frac{1-\alpha^2}{2\gamma}}$.

10.14 Higher θ implies faster mean reversion. Since $\alpha = e^{-\theta \Delta t}$, larger θ yields smaller α, meaning less persistence in state values.

10.15 From $\mu = \frac{\beta}{1-\alpha}$, we find $\beta = 5(1 - 0.8) = 1$.

10.16 Estimate parameters via: - Linear regression of X_{n+1} on X_n to get α and β, - Use residuals to compute σ_d as the standard deviation of $X_{n+1} - (\alpha X_n + \beta)$.

10.17

$$X_1 = 0.95 \cdot 0 + 1.5 + 0.2(-0.8) = 1.5 - 0.16 = 1.34$$
$$X_2 = 0.95 \cdot 1.34 + 1.5 + 0.2(0.5) = 1.273 + 1.5 + 0.1 = 2.873$$

10.18 - μ: carrying capacity, - θ: rate of return to capacity (birth-death balance), - σ: environmental fluctuations or random shocks to population.

10.19 As $\Delta t \to 0$:

$$\alpha = e^{-\theta\Delta t} \approx 1 - \theta\Delta t, \quad \beta = \mu(1-\alpha) \approx \mu\theta\Delta t, \quad \sigma_d^2 \approx \sigma^2\Delta t$$

10.20 For $|\alpha| < 1$: process is stationary, mean-reverting.
For $|\alpha| = 1$: it becomes a random walk with drift, which is non-stationary and diverges over time.

Index

M. Catak and T. Allahviranloo, *Discrete Random Processes*, Studies in Systems, Decision and Control 644, https://doi.org/10.1007/978-3-032-08630-3

MIX
Papier aus verantwortungsvollen Quellen
Paper from responsible sources
FSC® C105338

If you have any concerns about our products,
you can contact us on
ProductSafety@springernature.com

In case Publisher is established outside the EU,
the EU authorized representative is:
Springer Nature Customer Service Center GmbH
Europaplatz 3, 69115 Heidelberg, Germany

Printed by Libri Plureos GmbH
in Hamburg, Germany